Innovation and Financial Markets

Innovation between Risk and Reward Set

coordinated by
Bernard Guilhon and Sandra Montchaud

Volume 7

Innovation and Financial Markets

Christophe Dispas
Georges Kayanakis
Nicolas Servel
Ludmila Striukova

ISTE

WILEY

First published 2021 in Great Britain and the United States by ISTE Ltd and John Wiley & Sons, Inc.

ISTE Ltd
27-37 St George's Road
London SW19 4EU
UK

www.iste.co.uk

John Wiley & Sons, Inc.
111 River Street
Hoboken, NJ 07030
USA

www.wiley.com

Library of Congress Control Number: 2021930258

British Library Cataloguing-in-Publication Data
A CIP record for this book is available from the British Library
ISBN 978-1-78630-066-9

MIX
Paper from responsible sources
FSC® C013604

Contents

Foreword

I did not hesitate in agreeing to write a few words in the foreword for a magnificent book on innovation and financial markets.

This is a major issue in Europe, and not only in France, in the 21st century, which has seen the brainpower revolution follow the industrialization revolution, which has been going on for two centuries, replacing the 2000-year-old agricultural revolution that led to the sedentarization of humans after the harvest period.

Innovation is developing massively with numerous advances in higher education and the rise of negative reactions against the excessive extraction of natural resources and raw materials, which deplete the planet's resources and require ecology and economy to cooperate.

The issue of innovation and financial markets is all the more important as we are in old Europe, with good human resources, education, creativity and innovation, but lagging behind in the transition from ideas to products that ensure the creation of wealth and jobs.

The brainpower revolution that is developing is only achieved through a channel of complex operations that are not limited to ideas and projects, but encompass a whole series of operations that the book analyzes, with method and accuracy as the criteria being observed.

Before a "start-up" (Americanism) project led by a small team can allow entrepreneurs to enter financial markets, multiple steps must be taken.

The authors have analyzed the criteria and present them in their innovative book. The final bibliographical references of these pages demonstrate this.

A crucial remark: global competition means that a good idea and a good project must come to fruition quickly and be protected.

While the number of ideas and projects, thanks to the quality of training and research institutions, is generally excellent in Europe, it is not the same for the transition of entrepreneurs in later stages, when the project must result in a company that must have products that satisfy customers. There are many criteria that the authors have highlighted.

The steps explained are too numerous for me to describe. Their succession is partly logical, but it is good to specify them.

Contractors must, of course, be tenacious and meticulous, ensure that the needs of potential clients are met, and know what criteria apply.

What is certain is that start-ups are not likely to succeed, but each failure gives you a better understanding of the market. There is nothing unfortunate about that, and in the United States, the kingdom of successful start-ups, it is often said that it is better to finance a team that has had one or more failures!

I mention the United States, because it must be said that many European start-ups – and often the best ones – migrate there to find financing quickly, which leads to a very large number of the most talented entrepreneurs migrating to the United States for this purpose.

This is one of the reasons why this book is important, as it seems that 1 million European entrepreneurs have emigrated to the United States, leading to the strengthening of that country's technical domination over Europe. Little research has clarified this figure. However, it seems that the British are the most numerous, ahead of the Germans and the French.

I also suggest that following this book, we try to create in Sophia Antipolis – because why not in Valbonne or Vallauris, France – a training structure for start-ups based on the appendices built up over the years before the initial book, to demonstrate its importance. And the foreword would have had the significance of a foreword: proposing a future!

And to make this future a reality, why not give an annual prize to one of the contributors who has a major ecological impact? Impact, for example, in the field of hydrogen fuel, or the transformation of an innovation that would halve the energy consumption of old buildings with the industrialization of silica hydride, a crystalline solid that weighs 1% of the weight of water because it is 99% air, and which enables ultra-light and ultra-insulating concrete to be sprayed on walls?

In France, we also benefit from an important legal innovation: the creation of "*sociétés à mission*" companies, whose articles of association make it possible to avoid hostile takeover bids thanks to board members (staff, suppliers, customers, representatives of environmental priorities), other than financial shareholders. In this way, sustainability would be privileged over the dangerous "short-termist" views of some traders.

Pierre LAFFITTE
Honorary Senator
General Engineer at the *Corps des Mines*
Creator of Sophia Antipolis, the *Club des Technopoles*
and the *Club des Cités Côtières*
Creator of the *Conférence des Grandes Écoles*
January 2021

Introduction

From start-up to the stock exchange is a long, often complicated, journey which we have tried to popularize and systematize in this book. We have chosen to structure this book through the major stages of an entrepreneur's life. It is addressed to all those interested in the business world, with or without prior experience. Both entrepreneurs and students will find a practical and accessible guide to increase their chances of transforming an original idea into a profitable business.

In Chapter 1, we present, in an original way and based on the personal experience of Georges Kayanakis, the different stages from the initial idea to the initial public offering. The practical highlighting of these different stages constitutes an unprecedented contribution to the literature on the subject.

Chapter 2 focuses on the elements that foster innovation in a company. Three elements emerge: siphoning requests to produce an idea, having the required skills and having a focused organization.

Chapter 3 returns to the challenges of innovation for companies. It is first of all a survival issue – without innovation the company can quickly become obsolete and disappear. But innovation is also about improving and creating new products/services, having a distinction, and being a leader. It is also the importance of reducing production costs, entering new markets, or creating a competitive advantage in CSR and/or the GDPR.

In Chapters 4 and 5, we elaborate on the financial elements for the company. Chapter 4 covers the main elements of the financial evaluation of an innovation project. Chapter 5 discusses the different sources of financing for innovation.

Starting in Chapter 6, we move from the prism of the company to that of the financial markets: once a company is listed on the stock market, how innovation is valued by the financial markets. In this way, we review the main models for valuing financial assets and try to make them accessible in a practical way through the use of Microsoft Excel.

In Chapter 7, we move from equity markets to bond markets. We will attempt to answer the question of the impact of innovation on corporate credit risk.

Finally, in Chapter 8, we will deal with financial innovation itself. Do new financial products provide investors with additional opportunities for diversification or do they contribute to increasing the risk of financial asset portfolios?

Resolutely practical and pragmatic, this book nevertheless provides numerous references to professional and academic literature to make it possible for readers to dig deeper into their favorite subjects.

1

From the Idea to the Stock Market

1.1. Prologue

I set up a company, ASK, which aimed to be a world leader in contactless technology. I thought that my vision was unstoppable, but I could not help thinking that the contactless smart cards that I was about to develop would be a huge disappointment in terms of functionality. I could not convince myself, of course, but I could not despair either: I told myself that I was missing something and that I needed to think more about the properties of the conductive ink. I went over everything again to make sure that my specifications were well adapted to the ultimate goal: "To sell contactless smart cards in ISO format, no more expensive than a contact smart card, that works a few meters away from a reader." The solution I chose was perfectly in line with the famous "vision". As the card had to be very light in order to be inexpensive, I opted for a paper substrate (the photocopy kind, not expensive) containing an antenna made of conductive Ag ink applied via silk printing, a technique whose cost for large volumes is almost equal to that of the material. "Away we go", I told myself. "I'm sure it will work. But there will be some fine-tuning!" I am not yet at the stage of using conductive ink on paper! I am only at the stage where I use a small brush on an epoxy glass plate. It does not work!

Pierre, the employee I had just hired, appears quietly behind me. "Are you coming to eat?" he asks me. "It is eight o'clock." I hesitate for a moment and say, "I'd rather stay. You go alone. I mustn't interrupt myself." He doesn't go away. He doesn't want to let me down. Living far away from our families, we are used to eating together in the evening before going to our hotel rooms. "I'll stay a little longer, if you don't mind." I don't listen to him anymore; I look at two Ag ink loops. Two millimeters wide, uncontrolled thickness, but apparently sufficient. I take the binoculars and inspect them again. Maybe something is missing? A discontinuity?

No, nothing! And the skin effect? What is there to get from that? I am thinking about it. I do not see it. There is so much material that it should pass! But it does not! It does not pass! It is the ink that does not conduct well enough. I have 60 ohms, when I should have 3 or 4. For the third or fourth time, I take the specification of the ink again. The manufacturer does not mention 3 or 4 ohms. This is not normal: they are small silver grains embedded in epoxy. They are supposed to touch each other to ensure the continuity of electron flow. So, they are not touching each other as much as they should be! I immediately call Pierre and say, "Pass me the hammer, please." I put my epoxy glass plate on the ground on which the antenna is placed at 60 ohms and I delicately tapped its surface with the hammer, taking care not to damage it. When both loops are finished, I take the ohmmeter and measure: 12 ohms! That is good. The Ag grains have come closer together and there are more of them touching each other. I continue tapping with the hammer. The resistance does not change anymore! I understood: the polymerized epoxy prevents them from going further. The mechanical treatment must be done before the end of the polymerization… Pierre remained silent, as did I. I set us free by saying, "It's good. It will work. Give me some paper!" At full speed with my brush, I lay down a new antenna. I put it in the oven at 80 degrees and wait a few minutes. I confess that I am guessing a little bit. I take the antenna out on its paper and I tap it with my hammer. It flattens more easily. I stop and measure. 3 ohms! When we get out of our little room, the night and the Aix cold envelop us. The ground is very slippery, icy. We go directly to the Formula 1 hotel, where we have chosen to stay. We are not hungry anymore, but we are thirsty for the future…

Box 1.1. *Excerpt from notes by Georges Kayanakis, December 1998*

And the future will prove us right. Indeed, in the 2000s, thanks to ASK products, RATP (a French state-owned public transport operator and maintainer) launched one of the precursors of contactless payment in France: the Navigo pass.

This client, among others, enabled ASK, which was founded in 1997, to become the world's leading manufacturer of contactless microprocessor cards, contactless paper tickets and contactless paper radio frequency labels, in a decade. The initial specialization in the public transport sector was extended to other industries, for identification purposes and biometric passports. ASK became a tier-2 supplier for companies such as Gemalto and Oberthur and its technology equipped US biometric passports.

The growth of this company, which had a turnover of 0–32 million in 10 years, led to an introduction in the new market in 2014. The company merged with Paragon two years later, to form the company Paragon ID, to become a world leader in identification, with a turnover of more than 100 million euros at the time of writing.

Since then, contactless technology has become widespread. According to a study conducted by MasterCard at the end of 2018 in Europe, nearly one out of every two transactions in stores is now carried out using contactless payment[1]. In 2020, the Covid-19 crisis gave the start signal for its massive use. This new use is based on an invention, transformed into an innovation, which took several decades to spread in our society.

The process of developing a company through innovation can be, at least, in part, systematized. In fact, as part of a national study evaluating companies based on their innovative capacities, the *Collège des Bernardins*, the *Centre de Gestion Scientifique* of the *École des Mines de Paris*, under the direction of Armand Hatchuel[2], and the Fondation Sophia Antipolis, under the direction of Georges Kayanakis, conducted a study to explore and propose solutions.

Under the chairmanship of its creator, Pierre Laffitte, the Foundation Sophia Antipolis was mostly in charge of the "proposal of new criteria for the evaluation of innovation capacities in companies".

The aim of this bottom-up study was to define the criteria required for innovation. It was organized according to the following modalities:

– 30 companies forming a representative sample and operating on the territory of Sophia Antipolis, linking to the PACA region's competitiveness clusters who were called upon to give their opinions on the factors that favor innovation, based on their own experiences;

– representatives in the companies had to be founders or managers of innovative companies with a valid opinion on the subject due to past or present experiences;

1 MasterCard (2018). Près d'une transaction sur deux en sans contact. Press release.

2 Professor at the *Ecole des Mines de Paris* and member of the *Académie des Technologies*. Biography available at: http://www.mines-paristech.fr/Services/Annuaire/armand-hatchuel.

– the only question they were asked was, based on their own experience, what the most favorable criteria for innovation (organizational, sequential, technological, relational, financial, commercial, marketing, product/market, strategic, etc.) were and which would have had, according to them, the most influence on the success or failure of the company;

– a team of six people (all members of the *Jeune Chambre Économique Française*) was formed to conduct this study, which lasted more than a year.

This work complemented the knowledge and personal experience of Georges Kayanakis, who in Chapters 1 and 2 summarizes the whole of this book, both the nature and succession of the steps and the criteria that is favorable for innovation.

1.2. A long and rigorous path

1.2.1. *A graphic representation*

In the economic world, an idea can eventually lead to an IPO. Depending on the structure that carries the idea, the steps that need to be taken will be numerous and difficult. A start-up has to go through all the steps described in the staircase in Figure 1.1, in order to become a new company.

Future entrepreneurs start alone, equipped with their idea, differentiator, business model and personal funds (generally low). They have the double objective of making a prototype with their funds, and surviving.

They have just created a start-up. Destination: the stock exchange!

Figure 1.1 shows the steps involved in taking an idea and making it into a public company. The time function is not shown.

Apart from the first three "steps" that set out the initial conditions needed for a start-up to exist, the other steps are either:

– a mandatory threshold that needs to be crossed (e.g. conducting a beta test, adjusting a prototype, etc.);

– a condition that needs to be met to validate a change (e.g. verification of…; confirmation of…);

– the generation of key documents (business plans, strategies, etc.).

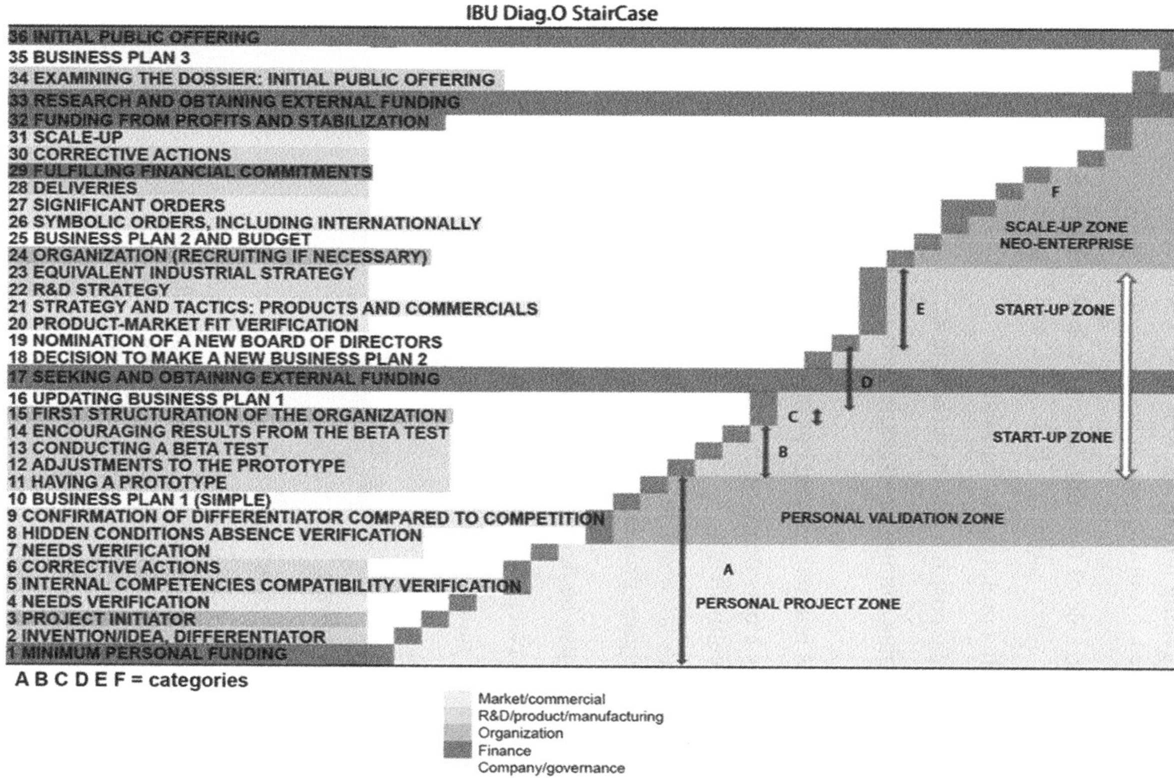

Figure 1.1. *Staircase (source: Georges Kayanakis). For a color version of this figure, see www.iste.co.uk/dispas/innovation.zip*

It should be noted that while the number of verification/confirmation steps is high at the beginning of the start-up (which corresponds to the natural concern of "getting it right the first time"), it becomes zero as soon as a beta test is launched. From then on, actions become predominant.

On the right hand side of the staircase, the conventional names of start-up milestones and their abbreviations A, B, C, D, E and F are indicated:

– A: start-up with a prototype or equivalent;

– B: *idem* A, plus a positive pilot test (beta test);

– C: *idem* B, plus minimal strategic neo-organization (founder and product marketing);

– D: *idem* C, plus internal, external or other funding;

– E: *idem* D, plus a new board of directors and market strategies (product, R&D, industrial or equivalent);

– F: *idem* E, plus a complete organization, written procedures, orders, clients, revenue, margins, profits, etc.

These steps can also be described in a more concise way:

– A: pre-start-up category: personal project and its validation;

– B, C, D, E: the four categories of start-ups;

– F: new business category.

1.2.2. *Steps to be taken*

1.2.2.1. *Steps 1, 2, 3: start-up*

These first critical steps must imperatively end with the development and in-house manufacturing of a product/system with one or more characteristics that differentiate it from its competitors. This phase is a personal phase during which the founder is the main actor, whatever the subject. In general, they provide the financing with their own funds.

Identifying and selecting the right opportunities for new ventures is one of the most important capabilities of a successful entrepreneur (Ardichvili *et al.* 2003).

1.2.2.2. *Step 4: scrupulous verification of the existence of needs*

This is one of the very first actions of a start-up that has just been formed. Time has passed and technology and competition are moving fast.

1.2.2.3. *Step 5: verification of compatibility with internal competencies*

Now that the needs of the identified market have been validated, it is necessary that these needs are converted into technical requirements that are compatible with the skills (or the DNA) of the company, and possibly with the subcontractors/suppliers.

1.2.2.4. *Step 6: corrective actions*

Depending on the differences between steps 4 and 5, corrective action must be taken to modify the product/service. Special care must be taken with the differentiator: it must always retain the intended importance.

This step is where the product/service–market fit must be carried out internally, and the validity of the differentiator must be confirmed before progressing further and investing significantly. In the absence of this fit, the product/service will not sell.

1.2.2.5. *Step 7: needs verification*

Young companies use a wide repertoire of behaviors in their attempts to meet new challenges. New companies usually do not have routines/procedures to deal with the many problems and opportunities they may face. Therefore, they take whatever measures they feel are promising, appropriate and available at the time. The result is a series of experiments: some planned, some unplanned and some accidental (Gong *et al.* 2005).

In any case, during this initial phase, the founders must check the product/service–market fit in the field they are looking at. As the product evolves between steps 4 and 7, it is imperative to make sure that these evolutions have not eliminated certain needs. During this step, it is entirely possible that a total fit is not obtained (performance, ergonomics, price, time-to-market, differentiator, etc.). In this case, it is necessary to go back to the market to see what conditions this product/service could be sold in. In general, a reduction in the selling price, or a reduction in the size of the market (going from TAM to SAM[3]), can be a compromise to satisfy the

3 TAM: Total Available Market; SAM: Salable Available Market or Target Market.

initial objectives. There must also be a reconsideration of the strategic and financial claims of the expected results. Unfortunately, it is very common to observe determination in a start-up which remains in its initial position. This can lead to a new innovation that solves the problem (which will require more financing and more time), or to the beginning of a decline, which, because of survival instinct, dramatically delays deadlines.

1.2.2.6. *Step 8: verification of the absence of hidden conditions*

If a compromise was necessary (this is the most frequent case), the product that was redefined is a different version of the product/service initially planned.

The situation can be summarized as follows: the start-up has a version of a product/service with a differentiator that can meet the needs of a perfectly defined market (target market).

Starting from this precise situation, it remains to be verified that there are no hidden conditions that prohibit, *de facto*, access to the target market.

By way of an example, we can refer to the case of an oligopolistic market in which a few competitors, in return for a stable market share, end up adopting the same high prices without trying to compete with each other (no price war). To maintain their market share, the arrival of a new contender will trigger a fall in prices among the competitors, *ipso facto*, thus modifying the value of the innovative investment.

1.2.2.7. *Step 9: confirmation of the differentiator's value*

This is the last preparatory step. It is a matter of verifying that, with all other things being equal, the differentiator, which has been altered (or modified) by the preceding steps, works in the intended market (target market). This verification must show that the product/service and its differentiator provide a "customer benefit" to the customer. Customers' choices are guided by their interests; one of the most important is the advantage that can be provided to them (financial, commercial, strategic, etc.).

1.2.2.8. *Step 10: business plan 1*

In recent years, there has been a debate about the importance of writing business plans, arguing that it is a waste of time and that it is better to devote time to the actual launch of the company. There are some articles that prove this opinion. However, it seems to us that formalizing, among other things, why the customers you are targeting are the highest potential target, why they will buy your products rather than those of your competitors and how you will succeed in selling them these products is a common-sense prerequisite to launching any business, rather than a waste of time. At the very least, writing a business plan will allow a founding team to explicitly confront its development hypotheses (sometimes unspoken), align its points of view, and then its resources, to work in the same direction. The vast majority of the ecosystem that accompanies business creators (incubators, accelerators, associations, funders, etc.) will ask for a business plan during their first interactions with project leaders.

Finally, we mention a significant study that was conducted on over 1,000 entrepreneurs, over six years in the United States, by two Harvard professors (Greene and Hopp 2017). It concludes that entrepreneurs who have written a business plan have a 16% greater chance of achieving profitability than those who have not done so.

Literature abounds on the content of business plans, but it is nothing more than explaining why you think your idea has real business potential. It is, therefore, a matter of covering all the fundamental topics that need to be addressed when launching a new business.

Regardless of the final structure of the plan, all four essential elements must be covered (Sahlman 1997):

– the team: the one that starts and develops the company, as well as members that are outsourced (lawyers, accountants, etc.);

– the opportunity: what will be sold and to whom, if it can develop and how fast, what is potentially on its way;

– context, overview: regulation, demographics, inflation, etc.;

– risk and reward: what can go wrong and what can go right, and how the team can respond.

When readers have finished reviewing the plans, they should have a good idea of what the entrepreneurs are trying to do (Dorf and Byers 2005).

It is interesting to note that simply formalizing a business plan and a request for funds significantly increases the chances of survival for young start-ups under 30 months of age (Delmar and Shane 2004).

Going into a little more detail, a business plan should cover specific points.

1.2.2.8.1. Vision/mission of the start-up

Based on its own DNA, the objective of the start-up is to indicate, in a simple way, what the company aims to become in the long-term and its mission within society.

1.2.2.8.2. Team

This is one of the most important points: how are the profiles and experiences of the entrepreneurs relevant to the project and why are their skills complementary?

Note that novice entrepreneurs who have little experience or references of their own that can help establish the cognitive legitimacy of their start-up may "borrow" this prestige (i.e. social rank or membership in exclusive social networks) from other team members, or from the advisory board of their new company (Mitteness *et al.* 2013).

This legitimacy helps start-ups to overcome their liabilities (Stinchcombe 1965) by strengthening the entrepreneurs' ability to create social ties and initiate routines to earn resources (Delmar and Shane 2004).

1.2.2.8.3. Problems or needs of potential clients

This is a description of the problem that the start-up/new product will solve. If the product is crucial to targeted customers, then the launched product will become a "must have", and customers will become captive.

Note that customer needs may be existing/expressed or latent.

In the process of progress (cost reduction, market share gain, quality improvement, increase in production capacity, improvement of margins, etc.), a need always expresses a lack, or a future lack, whether it is downstream, close to the end use, such as the empowerment of individuals, or upstream, the improvement of knowledge in fundamental research. This lack has two aspects: one is obvious (or existing) and the other is yet to be discovered (or latent). The existing needs, for those who express them, reflect gaps of different natures requiring certain actions:

– find a supplier for a machine, equipment, service, know-how, etc.;

– find a supplier for a new product;

– create a beneficial competitive offer (price, deadlines, quality, volume).

Latent needs, on the other hand, must be discovered (or deduced), because if, in general, those with this potential are aware of the imperative to "move forward", they are most often unable, for lack of the means to analyze all of the offers in detail, to specify the outline of what they would like. The difficulty for those interested in serving these potential markets is to reconstitute, from the outside, the environment of their future customers and to imagine what their own DNA can bring as a decisive customer benefit.

Focusing on existing needs (client-centered approach) is generally short-term and reactive in nature. A market-driven approach, which goes beyond meeting expressed needs and aims to understand and satisfy latent customer needs, is long-term and proactive in nature (Slater and Narver 1998).

Careful study of market needs can help an entrepreneur begin to develop an opportunity, but may or may not result in the creation of a business (Ardichvili *et al.* 2003). Thus, this is only the beginning of the process.

1.2.2.8.4. Description of the product

The description of the proposed solution must be simple and not contain the technical/intrinsic characteristics of the product, but must especially contain the benefits that this solution will bring to customers.

In this step, the initiator will also be able to explain how their solution is better than the existing alternatives on the market.

1.2.2.8.5. Intended market/target

This is an obvious fact, but one that is worth remembering, as some entrepreneurs forget: a founder will have a better chance of building a company of a significant size if they are operating in a growing market of a respectable size, than if they are targeting a declining micro-market.

Within the targeted market, the start-up will have to identify its priority segments. A start-up must prioritize these. Whether they are based on existing or latent needs, the market must be segmented along two axes: links with the DNA and time to market (see Figure 1.2).

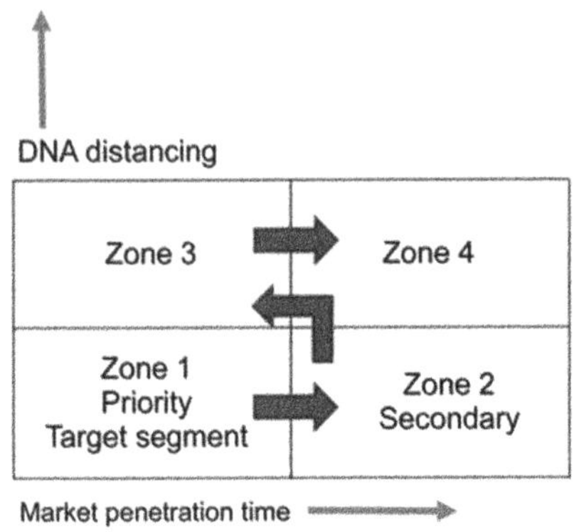

Zone 1: priority zone. This is the target segment.

Zone 2: secondary zone. It is a direct potential to be addressed as soon as the company has the means to do so without disturbing its achievements obtained in zone 1.

Zones 3 and 4: in order, subsequent growth drivers.

Figure 1.2. *Target markets as a function of DNA. For a color version of this figure, see www.iste.co.uk/dispas/innovation.zip*

1.2.2.8.6. Business model

The business model must demonstrate how the start-up will quickly generate revenue by serving addressable market segments and, in the longer term, make a profit, i.e. to succeed in producing its products/services at lower costs than customers are willing to pay for them.

– Depending on the target, a business model must be developed based on constraints and answers to fundamental questions:

- what qualifications/certification clauses apply? For example, to be able to sell equipment in the defense industry, the equipment must comply with the standard specifications in terms of mechanical, climatic and electrical constraints;

- is the written specification of the product/service really feasible? If so, what investments need to be made?

- the market price (the answer to the question: at what price can we sell?);

- what are the monetary clauses? What are the exchange rate guarantees and what are the costs for the guarantees?

- delivery periods (to answer the question: how long between order and delivery?);

- what are the penalties for delay?

- what are the quality, reliability and documentation clauses and their organizational and financial consequences?

- what are the required guarantees and the cost of the insurance being taken out?

– Stakeholders: there is no business model that does not address the added value that guarantees a certain durability.

How is what is sold unique and indispensable to the customer? This is one of the fundamental questions.

Regardless of the danger created by the existence of competitors, the target market may simply change or even disappear due to the evolution of the client's strategy. The business model must address its sustainability, i.e. the inclusion of its added value in the stakeholders' value.

With respect to business models, there are several widely adopted frameworks, like the *Osterwalder and Pigneur's Business Model Canvas,* which allow the evaluation of a particular business model at a given point in time. But there are also frameworks, like the *360° Business Model Innovation* (Rayna and Striukova 2016a), which allow entrepreneurs to have a more dynamic picture.

According to Rayna and Striukova (2016a), the business model must demonstrate how the value of the company will be:

– created, one of the most important elements (Zott and Amit 2002; Chesbrough 2007);

– proposed, what is being offered (the product or service), and at what price (the pricing model) (Chesbrough 2010; Teece 2010, 2011);

– delivered, market segments/target distribution channels (Osterwalder *et al.* 2005; Abdelkafi *et al.* 2013);

– captured, refers to the firm's ability to benefit from the value created (Chesbrough 2007; Holm *et al.* 2013);

– communicated to clients and partners (Bieger and Reinhold 2011; Abdelkafi *et al.* 2013).

Few aspects of new business creation attract as much attention as the business plan. Numerous articles have been written on the subject and numerous competitions have been held around the world.

1.2.2.8.7. Competition

It is the group composed of suppliers who can directly or indirectly satisfy existing or latent needs:

– multifaceted competition: very generally, the competition is made up of companies that can be easily identified through a well-adapted "business" filter. But it is increasingly common to find that certain related businesses (those that have a connection with the target in other aspects) "encompass" the product in a broader offer, thus becoming competitors not on the product, but on the system that encompasses it. These "related" professions can be very different from one another. Often, these offerings are more attractive to the end customer for financial saving reasons;

– competition from nowhere: it is globalization and the power of modern communication that shortens distances and reduces time. It is not at all uncommon to discover, during a call for tenders, that there are one or more competitors, very valid, located in a distant country, out of reach of the usual radars.

1.2.2.8.8. Commercial and production strategy

The business plan should describe how the company will reach its target customers, including the promotions, channels, partners and tools it will use to do so.

In a more detailed and concrete manner, the start-up must clearly identify the targeted customers as well as their planned orders, described and quantified (in volume and price) month by month (or quarter by quarter).

If the start-up is lucky enough to make some sales at the time of writing the business plan, this will give it a lot of credibility. Of course, the forecast figures will have to be in line with the actual metrics.

The forecast must represent the needs of the market regardless of the possibility of serving it. However, a judgment must be made on incoming orders in order to take into account the probability of manufacturing and thus give rise to sales forecasts.

1.2.2.8.9. Financial forecasting

Spread over the same periods as the sales plan, a revenue plan derived from sales forecasts will be drawn up. It includes the following lines: market volume, sales volume, sales price, cost of sales and gross margin.

The resources that will be put in place to achieve the planned figures must be identified. These are usually made up of the following items: promotional activities, expenses, purchases, human resources, R&D, etc.

A manufacturing cost forecast must be prepared for the same period. These costs must reflect the various measures taken to reduce costs in order to follow the evolution of selling prices. These costs are broken down as follows: materials, labor, subcontracting, and production expenses. Purchases of equipment are identified, costed and placed at the time they occur.

All of these elements will make it possible to build an operating and cash flow forecast account. They also give the value of certain ratios such as ROI and payback. Whether it is for a start-up or for a project in a well-established company, this plan must be followed even if (as is the case in a start-up) some information is difficult to establish.

1.2.2.9. *Step 11: having a prototype*

Once the product/service is well defined, it must be transformed into a real product: the prototype. This step, which is very technical, must be focused on the most faithful realization of the planned functionalities described at the end of step 9. During this step, particular care must be taken to ensure that the differentiator is implemented.

However, in terms of efficiency, and at short notice, the search for a prototype centered on the essential, but imperfect, is far preferable to that of the theoretical perfection, which causes dangerous delays and uncontrollable consumption of resources, while being exposed to risky results. Delivering an imperfect prototype, which performs the essential functions, to the customer's premises as quickly as possible makes it possible to quickly reach a satisfactory development for both parties through successive iterations.

1.2.2.10. *Step 12: prototype adjustments*

The performance of the prototype built in step 11 must be formally verified against the initial specification. During this operation, the customer's behavior and environment must be simulated, as in normal operational conditions, in the presence, if possible, of members of the start-up company that directly produced the prototype. It is very rare not to notice a certain number of defects. This necessarily leads to minor or major adjustments. In any case, the consequences must be evaluated and measured.

These adjustments are rarely impossible to make. The most important of them, on the other hand, can lead to adverse consequences such as: longer uptime, increased costs, and changes in expected performance. In these cases, a strategic decision must be made.

In this respect, the use of 3D printing within the company, or on online platforms, makes it possible to develop prototypes and reduce waiting time (Rayna and Striukova 2016b).

1.2.2.11. *Step 13: conducting a beta test*

From the first to the 12th step, the person in charge of product marketing (often the founder in a start-up) has identified, in the target market, one or two potential customers willing to conduct an evaluation of the prototype in real conditions. This is the "beta test".

For this test to be effective, a protocol must be established beforehand, stipulating that:

– access to information, resulting from use, and the possibility of carrying out corrective actions during testing are guaranteed;

– the customer, at the end of the test, will write a critical report on the achievement of their initial objectives;

– the customer authorizes the start-up to use its reference on the market.

1.2.2.12. *Step 14: encouraging beta test results*

The beta test is a cardinal operation. For the duration of the test, there must be a permanent relationship between the customer and the supplier. The start-up or project must be reactive to any problems that could arise from the use of the product/service in real conditions. At the end of the beta test, the product/service must have responded to as many criticisms as possible.

At the end of this step, the customer will give their opinion on the product/service, with regard to the objectives stipulated at the beginning, in a report. This report can be positive even if it contains some restrictions. Indeed, depending on the nature of these restrictions, we can judge that the overall results are encouraging. If not, a return to step 5 is necessary.

1.2.2.13. *Step 15: initial structuring of the organization*

If the beta test results are encouraging, there is good reason to believe that the actual product/service is a good fit for the intended market. Therefore, we must begin to create a structure to serve it by having an organization with at least the following five roles:

– management;

– R&D;

– product marketing;

– production;

– sales.

For obvious reasons of complementary, but not identical, objectives some roles cannot be cumulated with others. This is the case for product marketing, which cannot be combined with any other role – especially with R&D or sales, as is often seen.

In earlier periods of the start-up, several of these roles could, for a certain period of time, be accumulated by the founder. In this step, this facility should be banned. It is necessary to individualize these roles as much as possible, with the awareness that product marketing will require a full-time professional reporting only to the manager, until a new organization guaranteeing the influence of this role on the manager is put into place later on.

Vesper (1990) stated that the process of creating an organization involves the acquisition of five key ingredients:

– the technical know-how to generate the company's product or service;

– the idea of the product or service that guides the organization's efforts;

– personal contacts, "because companies are not launched in isolation";

– material resources;

– customer orders.

1.2.2.14. *Step 16: updating business plan 1*

As soon as the beta test shows encouraging results, the expectations contained in the first business plan should be updated by integrating the impact of the new organization on costs, revenue, profit and also the impact of the remarks of the customers who welcomed the beta test.

1.2.2.15. *Step 17: seeking and obtaining external funding*

The start-up – which now has a product/service qualified by one or more positive beta tests and an accessible market – can seek external funding, which is most often modest, in order to develop.

It should be noted, in addition to the lack of attractiveness caused by the small amounts sought, that in this step of the start-up's development, investment funds (very important sources of funding) are unlikely to have interest, as the company is still too young to be able to offer exit strategies, which, in this case, must go hand in hand with entry.

These modest amounts are of interest to investors such as business angels, interest-free loan networks and crowdfunding, as detailed in Chapter 5.

In any case, funders rely primarily on what already exists to justify their decisions. Existing data is made up of financial data and elements that can complement and possibly reinforce them, such as:

– vision;

– potential of the addressed market;

– potential of the technology used (growth reserves);

– relevance of differentiators;

– protection against competition (patent, leadership, etc.);

– recent and tangible proof of product–market fit (existence of a prototype and a positive beta test carried out with one or more customers);

– scalability;

– HR quality (especially those of the founder at this stage);

– complementary organization;

– ability to transform an idea into an innovation;

– procedures;

– risks, etc.

This complementary part, which is essential to justify the forecasts, is an "invisible asset" of the start-up. To express all of its importance, this immaterial force requires that all the information constituting it be organized, in order to make a coherent and "digestible" whole.

This action is all the more difficult to achieve because the start-up may not be aware of its importance as per the radical adage that the numbers indirectly reflect ... So, we work on the numbers, but not enough for their justification!

1.2.2.16. *Steps 18 and 19: decision to make a new business plan and appointment of a new board of directors*

Although distinct, the two steps of deciding on a business plan 2 and setting up a new board of directors are usually initiated simultaneously.

1.2.2.16.1. Step 18: decision to build the business plan 2

If the first investment has taken place, it is necessary to make a new business plan (2) whose steps, from 19 to 21, will constitute the "steps sequence", and steps 23 and 24, the consequences in organizational and financial terms.

It will be based on the precedent, modified by the various requests and criticisms that have emerged during the negotiations with investors. The investors expect that their contributions and the different requests will be integrated into the budget and consolidated with the figures that have attracted them. The latest information from the market, which would have been pouring in since the initial business plan was published, must also be integrated into the new plan.

It is also the time to foresee, through this new plan, the filling of some organizational gaps coming from the previous lack of means (see step 15). This is the period when compromises abound and the restrictive but imperative notion of intermediate results (milestones) appears. This notion crystallizes the will to respect the plan despite difficulties. As soon as a balance is found, intermediate results serve as fixed benchmarks along the way to achieving the objectives.

1.2.2.16.2. Step 19: modification of the board of directors (the board)

Following the investment made by external persons (individuals or legal entities), it is natural, in parallel to the decision to make a new business plan, to modify the board of directors to include one (or more) of the investor's representatives. This is the moment when the notions of company – a non-legal entity grouping together means – and company – a legal entity focused on the beneficiaries – dissociate.

Although this step may seem benign in terms of its impact on the development of the start-up, it is nonetheless very difficult to manage: the minimum amount of formalism required (reporting, regular board meetings, etc.) diverts a significant amount of the founder's energy from their main tasks and skills. A new balance has to be found between the various obligations arising from having new shareholders.

1.2.2.17. *Steps 20, 21, 22, 23, 24, 25: verification of the product/system fit for the market; development of all strategies and tactics (commercial, R&D, industrial); hiring*

1.2.2.17.1. Preliminaries

One of the most common elements shared by war and entrepreneurship is a lack of certainty. Resources, reactions and the environment all contribute to making war a virtually unpredictable soup of ingredients (Honig 2004). Therefore, it is essential to constantly rethink these "ingredients".

The very long period (from one to several years) covering steps 16–20 can be seen from various angles:

– the start-up, which was acting according to business plan 1 and for which it was imperative to exploit its assets as a priority and measure their effectiveness. To do so, this period was characterized by a willingness to not change the parameters for a certain period of time (in any case, long enough to get significant feedback from the markets);

– the competition, for whom a long period of time was necessary to put in place the corrective actions intended to react (or improve their positions) to the new offer;

– the market, for whom the integration and profitability of products/services from a supplier of generation n requires a fairly long time, most likely several years, which makes it unlikely that a generation $n + 1$ will replace the first one too quickly (it is extremely rare to replace generation n with generation $n + 1$ without any benefit).

During this long period, where apparently nothing happens, many changes (price, features, performance, ergonomics, consumption, etc.) may be in gestation. Anticipating them is not only an advantage, but is above all a necessity.

A contingency plan can also be drafted during this step. This plan must be adapted to the needs of the company and must allow for the following three objectives to be achieved (Iaquinto and Spinelli 2006):

– provide emotional, intellectual and financial support in the aftermath of a potential business failure;

– provide a base from which to launch the next business;

– facilitate reintegration into the market.

1.2.2.17.2. New bases for business plan 2

Steps 20–25 are key. They determine the future company. Steps 20 and 21 can be described as a transition between the start-up and the future company.

As the first steps of the future company, they will constitute business plan 2, which will have to be built on the following bases (different from those of business plan 1):

– take into account that the start-up, which started from a simple idea, has achieved a significant result. It has, therefore, acquired innovation capacities;

– to not jeopardize the main assets of the start-up, soon to be a scale-up/neo-enterprise, which are:

- the existence of products/services, even if imperfect, that have been sold,

- the existence of a clientele,

- the existence of an image, however modest it may be,

- a know-how, patents perhaps,

- adapted HR,

- initial revenue, etc.;

– evolve rather than revolutionize: on the basis of the initial plan, adjust, modify, add, subtract, etc. The most urgent issue is always the product–market fit, which must be the main focus of the new business plan;

– set strategic objectives that are accessible (revenue, market penetration, profitability) with a high probability of success and milestones;

– respect financial constraints: the plan, including the budget, must be presented according to two assumptions: conservative and aggressive. The first incorporates strict adherence to financial possibilities, while proposing reasonable objectives. The second, on the other hand, proposes more aggressive, less certain and costlier objectives. If the second hypothesis is retained by management, the additional costs and benefits will have to be duly identified and isolated in order to be able to judge the parameter: potential gains/risks.

1.2.2.17.3. Business plan 2, its strategic role and content

This new business plan is fundamental for the following two reasons:

– it can be described as a "basic strategic plan", since its main elements will eventually be included in an IPO or capital increase;

– if it builds on the foundations outlined in the previous section, it is likely to foster a culture of success (procedures, know-how, positive market knowledge, etc.), which is the key to building an effective and repeatable processes.

As for any business plan, it must include:

– checking the product/service–market fit (see step 23);

– product and commercial strategy (see step 24);

– R&D strategy (see step 25);

– industrial or production strategy (see step 26).

And for the phases of consequences:

– organization, HR hiring;

– financial plan (operating account, investments, working capital, cash flow, balance sheet).

1.2.2.17.4. Business plan 2, strategic plan, its construction

This is elaborated on in steps 20–25.

1.2.2.17.5. Step 20: verification of the product/service–market fit

Product marketing, within the scope of its roles and according to appropriate methods, gives rise to new technical specifications and defines new price targets. It verifies the product/service–market fit and, if necessary, identifies the corrective actions that need to be implemented. Their importance may range from minor modifications to the production and launch of version two of the product/service, for which they will also check the product/service–market fit. This development phase can be accompanied by an appropriate commercial strategy (e.g. a more attractive pricing policy).

1.2.2.17.6. Step 21: establishing strategies and tactics for the product/service and for the salesperson

Having a commercial strategy

It first involves setting up a process that allows, in the target market, the conversion of potential customers into actual customers of the product/service.

Each step in this process must end with a partial result necessary to move on to the next step:

– identify potential customers among foreigners (identify);

– contact potential customers (connect);

– confirm that these potential clients have needs that can be served (explore);

– check with clients that the product is suitable for them: it is an opportunity (advise).

Getting an order (customer)

To achieve each of the partial results, tactics (tools in the broadest sense) must be developed.

Their nature can range from a simple phone call to a high-performance website, through a CRM, the purchase of databases, the establishment of marketing channels, etc.

The business strategy must detail all these tools, without forgetting, of course, the organization and the necessary human resources.

Having a product strategy and tactics

This is an essential "upstream" condition to achieve the start-up's objectives. Answering the question "which approach, how" (strategy) and "with what tools" (tactics) to reach the final objectives is the object of a product strategy.

The term "strategy" must be broken down to better inscribe it upstream in the DNA of the company, and downstream in the present and potential needs of the market. To do so, the product marketing must clearly address and feed the following points:

– mission of the company: why does it exist?[4]

– what is its DNA?

– guiding values;

– description of the product–service, its possible trajectories, differentiators and cost objectives to support sales prices;

– comparative table with the competition/benchmark;

4 After the entry of new shareholders into a start-up, it is recommended that these points are clarified so as not to leave room for interpretations, which could later cause problems.

– objectives: what targets? What are the target figures in terms of revenue, sales price, volumes, gross margin, penetration? What is the timetable?

– strategy: what approach should be used for each target and each objective?

– tactics: what tools should be used to implement the strategy?

– what resources are required (HR, promotional campaigns, digital marketing, exhibition, conference, etc.)?

– key performance indicators (KPIs).

1.2.2.17.7. Step 22: R&D strategy

In the short- and medium-term, it is directly deduced from the product marketing recommendations:

– specification of the product/service and its future trajectories;

– cost levels required to support recommended prices.

The objectives of R&D are:

– to develop the products/services and the associated differentiator(s);

– to define the road map in which the products/services fit (at least two future generations in order to anticipate market needs);

– to integrate cost objectives (basis of the selling price) from the design phase;

– in the long-term, it is important for R&D to look for innovative solutions. To this end, it must not only be constantly informed of what is happening in its technical field and on the market, but, in addition, it must isolate a human resource within its organization to be specifically dedicated to this type of role.

In the case of a start-up, as early as the 15th step where external funding is sought, it is important not to omit (which is unfortunately too often the case), to plan R&D resources and their funding, to be dedicated to the long-term.

Assessing R&D capacity can simplify the future implementation of the commercialization strategy; developing R&D capacity in the future can also help maintain competitive advantages (Lin *et al.* 2011).

1.2.2.17.8. Step 23: industrial strategy or equivalent

It is based on volume figures spread over time for the products/services provided by the product marketing (see step 20). These volume figures are often an estimate of customer needs. This estimate leaves room for uncertainties that contradict the commitment to supply customers on time.

The industrial or equivalent strategy, depending on whether it concerns products or services, has the objective of fulfilling delivery commitments in time, hours and quality to customers, despite any difficulties that may be encountered. Another facet of this strategy is to secure the company's position by maintaining concrete advantages over its competitors.

For this purpose, based on projected volumes and costs, the following topics will have to be analyzed, and answers must be provided:

– establishment of forecasted costs (detailing variable costs and fixed costs), whose objectives were given by the product marketing. If these objectives could not be achieved, then successive cost reduction programs must be planned;

– what capacity (in volume) to install?

– what investments, tangible (equipment) and intangible (training)?

– where to produce (sourcing): in-house, subcontracting?

– how to guarantee quality and reliability?

– production planning;

– strategic inventories and stocks;

– policy of supply and control of components or subcontracted parts;

– organization;

– human resources.

1.2.2.17.9. Step 24: organization

The different strategies, all of which project the former start-up towards the future, towards becoming a company, call for resources, the main ones being: HR, expenses, investments, as well as organization. This is the time to adopt the structure of a company that has the duty to be sustainable and profitable. The budget forecasts that will be the crowning achievement of these steps (from 20 to 25) will have to incorporate the expenses associated with this new organization.

The management of the start-up generally delegates HR responsibilities to a single person. They have to build everything from scratch, but that also means they can have an immediate impact. The mentality is similar to that of product development: deliver the solution, run it, then release version 2.0 if it fails, or if it is not perfect (Rogers and Paul 2018).

1.2.2.17.10. Step 25: business plan 2 and budget forecast

Once the forecast of revenue, volumes and sales prices has been established; recruitment has been scheduled; and costs have been planned and timed; the nature and amounts of investments (as well as the rules for their depreciation) have been scheduled throughout the three years; the standard costs to be used to value work-in-progress and in stock have been defined, etc.; the company is then in possession of the basic strategic figures that can enable it to draw up three-year budget forecasts. These will take the classic form: operating account, working capital, cash flow and balance sheets. They will be detailed month by month for the first year, quarter by quarter for the second and in full for the third.

These budget forecasts complement business plan 2. This will now be the reference platform for the new company.

1.2.2.18. *Steps 26 and 27: symbolic and significant orders*

The turbulent period surrounding the arrival of an investor has come to an end. During this period, the sale of the old products/services was continued, while awaiting the availability of the new versions that resulted from their evolution.

At the end of this turbulent period, the young company has updated all the elements constituting its assets:

– the products/services have been, or are being adapted to fulfill needs;

– these same needs have also been updated;

– the organization has been completed;

– the investors' funds have made it possible to fill in the gaps and give credibility to expectations;

– business plan 2, shared by all players, stakeholders and rights holders, is a launch pad for the future.

As this platform differs from that of business plan 1 (two major changes: products/services and market needs), it becomes necessary to verify that these changes are indeed effective, by obtaining symbolic orders for the new version of products/services from customers who are in the process of changing suppliers.

At the same time, for customers who are still loyal, but who have the same potential problem, it is normal to obtain (after a qualification phase) a significant order for the outright replacement of the old version by the new one.

At the end of these two steps, the company can observe that the product/service–market fit is achieved on a theoretical level.

1.2.2.19. *Step 28: delivery of the new version*

The theoretical product/service–market fit must be completed by the delivery according to the order: quantity (for industrial products), performance, quality, price, time.

If these conditions are met, the company can be considered an expert in its products/services. All the roles involved have achieved their objectives: product marketing, sales, R&D, production.

At the end of step 27, the product/service–market fit is reached, constituting the major asset of a company. However, in order to lead it, one day, to an IPO, this asset must lead to financial results in line with budget forecasts, the only valid contract between company and company.

1.2.2.20. *Step 29: compliance with financial commitments*

The business plan 2 generated dashboards for financial monitoring as well as milestones – meeting points – to take stock. These meetings are:

– daily (each manager has their own dashboard and/or procedure);

– monthly (actual operating statement for the past month and forecast for the next three months);

– quarterly (complete actual and forecast financial results for the remainder of the year);

– annually (actual full financial results).

These different means of monitoring the evolution of "key" parameters in near real time allow for micro-adjustments to avoid drifts, and also serve as an alert. If the results are partially or totally insufficient, certain corrective actions are necessary. They become imperative and a priority if they are important. Most often, in such cases, investments and recruitments must be suspended while awaiting the results of these actions.

1.2.2.21. *Step 30: corrective actions*

When the financial results are not in line with forecasts, one or more cost items are certainly in question: manufacturing costs, operating expenditures, depreciation and amortization. If the analysis shows that their potential recurrence is unavoidable (e.g. a gross margin problem due to the appearance of a new competitor producing at a lower cost), it becomes imperative to make a number of decisions:

– downward budget revision;

– reduction of hiring or hiring freeze;

– change in delivery from suppliers and subcontractors;

– decrease in operating expenses (travel, exhibitions, equipment purchases, etc.);

– postponement, or even freezing, of certain investments;

– immediate commitment of specific programs to reduce manufacturing costs without altering functionality, differentiators or quality/reliability.

It is recommended that the company have an ongoing activity to reduce costs and expenses.

1.2.2.22. *Step 31: scale-up*

The company has just completed a complete cycle since the entry of investors. The products/services have been modified to better fit the market demands, the financial results have been analyzed and corrective actions have been taken to solve the problems.

This cycle demonstrates the company's mastery of all the aspects required in launching a new product or new activity, and more particularly, the indispensable ability to react that must accompany any budget forecast which, by its very nature, cannot foresee everything.

The company is therefore ready to launch. The products/services, thanks to an appropriate communication policy, are made available to the entire potential market.

The resources used to accomplish this first cycle can be gradually devoted to other projects (new products/services), and procedures (type: routine) are put in place to monitor key parameters such as: the competition's activity, prices, technology evolution, quality, reliability, customer satisfaction, continuous improvement, etc., as well as internally, order entry, operating results, inventive activity, continuous improvement, etc.

These procedures will be a part of the intangible wealth of the company. They can allow the start of new projects, at any moment, without having to learn everything over again.

During these procedures, entrepreneurs consider the impact of growth on the company's profits – both economic and non-economic. Entrepreneurs also consider the impact that growth will have on the company's risks – both in terms of the company's ability to survive and the ability to maintain the quality of its products and services. A third, much less important consideration is whether it will require them to spend less time on their favorite tasks, as well as the extent to which growth will increase their total workload (Valliere 2006).

1.2.2.23. *Step 32: profit financing and stabilization*

The management team, which has been gradually put in place, has the tools it needs to drive the company's growth, while pursuing an essential goal: being profitable and self-financing its needs.

For the former start-up, the financial results must be judged on the basis of ratios that will be compared to those of a similar company. The management must become familiar with the handling of all the financial ratios used, which will be very useful for them later on, especially during an IPO.

Ensuring the financing of its needs through its profits has four important virtues:

– motivating staff through pride in being profitable (recognition of merit);

– not spending precious energy finding (or not finding) funders;

– avoiding costly financial expenses;

– attracting investors and facilitating IPOs.

This step is also that of stabilization, which can be short (one to two years) or long (up to many decades). The new company, which has become a normal business, continues to develop and grow, depending on its financial possibilities.

1.2.2.24. *Step 33: scale-up, VC (venture capital) financing*

As the scale-up continues, higher growth opportunities emerge, well in line with the company's capabilities. In order to seize these opportunities, the company must often consider leaving its financial comfort zone and taking risks that can prove fatal. In order to not jeopardize its sustainability, financing is sought to support these superior growth opportunities. The players that seem to be best placed to support this phase are venture capital firms, which help companies generate a much greater impact on innovation than other means of financing (Dutta and Foltab 2016).

Thus, more funds are needed to proceed with this acceleration and strengthen its equity and teams. Candidate funders with an interest in this sector of activity, and possibly in this/these technology(ies), are often concerned about the profitability that could be achieved with the funds invested. To this end, they not only rely on the actual financial results, but also on the updated business plan 2, supplemented by a "potential plus" section, identifying as precisely as possible the beneficial impact that this increase in revenue would have on the financial results over a period of at least three years.

In addition to these purely financial aspects, there is also a "judgment", the purpose of which is to convince ourselves that the company has the capacity to manage this type of major shift and to innovate (intangible assets) to ensure profits, growth and sustainability. It should be noted that the relevant information is often contained in the data room.

This last consideration – an important part of goodwill – is cardinal, as it can swing the decision to participate in the funding to one side or the other.

One of the preferential exit mechanisms for venture capitalists is the IPO (De Clercq *et al.* 2006).

1.2.2.25. *Step 34: processing of the IPO file*

If the results and prospects are excellent, with the agreement and collaboration of the shareholders, the company's manager and CFO can devote some resources to learn the conditions required to be listed on the stock exchange. In particular, they will be able to identify the nature and intensity of the efforts that need to be made, in parallel with the normal management of the company, which often constitute a problem that needs to be solved in order to carry out this operation.

1.2.2.26. *Step 35: business plan 3*

With the newly acquired funding, and as for the steps from 17 to 24 (business plan 2), a new business plan must be drawn up in order to "set in stone" the commitments made by the various players in the capital increase, and to take into account changes in the market and competition that may have taken place during this long negotiation period.

After the acquisition of this new financial capacity, the company must achieve the objectives set out in business plan 3, with a focus on results and the preservation, or even increase, of its innovative power, a major weapon in the context of an open market, where competition is global, and technologies succeed one another at a sustained pace.

This period can last for many years if the business model chosen involves controlled, self-financed, profitable growth, rather than risk-taking that generates strong growth but requires significant new funds.

1.2.2.27. *Step 36: IPO*

Strong growth must be properly financed without endangering the company's health. In view of the characteristics of the company and its performance, its potential to attract institutional or individual investors is such that its listing on the stock exchange is conceivable.

This has four major merits:

– at the time of the IPO, shares may be issued or sold and thus constitute a new source of funding growth;

– the sale of the shares allows the first investors to exit if they want to;

– individual holders (natural persons) of shares may, if they wish to, sell shares;

– issue stock options to key personnel with a view on an IPO.

From this point of view, an IPO seems to be the most appropriate solution. But, as we will see in Chapter 5, the decision about going public must not only take into account the firm's ability to continue this activity, but also a strategic assessment of how to maximize the value of the firm (Bruton and Prasad 1997).

1.2.3. *Product marketing by the menu*

It all starts with a person who has an idea, an invention or some available knowledge that they want to transform into an innovation.

Let us recall in passing that an innovation is an invention that sells profitably.

1.2.3.1. *Founding actions and the first product step*

For this, the future innovator must spend money and energy.

They make arrangements, form a company, perhaps hire one or more collaborators, and then, the first product step that needs to be taken presents itself.

What is the definition of the functional scope of the product: what should it do? What are the functions it must perform? What are the specific characteristics?

1.2.3.2. *Functional scope of the product or technical specifications*

The innovator must first get as clear an idea as possible of what will be purchased. In other words, they must anticipate the needs of the market.

In general, they already have their idea. But, in this step, the idea has to be consolidated by other means, means that are preferably easily accessible to him/her. This last point is cardinal since the entrepreneur usually has very limited resources and time.

At the very least, the entrepreneur must search in relevant documents, the Internet and in the opinions of specialists, for information that is independent and as unquestionable as possible.

Then, they have to write a technical specification (functional scope), since they are now in possession of a confirmed vision and sometimes even a conviction, on:

– the product they want to develop;

– a certain market that they believe will be in demand.

They can now specify, in writing, the broad outlines of their product in terms of features, price and other important parameters. At this stage, they think they know which product to develop.

1.2.3.3. *Prototype*

After many efforts and difficulties, the day finally arrives when they have a qualified prototype, carrying their basic idea, which should be sold on the market.

1.2.3.4. *Beta test, its usefulness*

Before thinking about selling the product. Proof that it provides the expected service in a "customer" environment must be obtained. This is the beta test.

1.2.3.5. *Beta test, the beginning of an iterative cycle*

First adjustments, product–market fit: the phase of searching for customers (market) who are willing to buy their product begins. It explicitly contains the two cardinal terms of the product marketing role – product and market – and is characterized by a desire for practicality.

An iterative supply-need adjustment cycle begins.

In this step, the prototype specification needs to be adapted to encourage a few customers to agree to host a beta test.

Who is in charge of the first product–market fit action? Up until now, and in general, the entrepreneur has spent most of their time on technically achieving their product. They put a lot of care into it. But did they do the same with the future acceptance of their product by the target market? In general, the answer is: no, because the entrepreneur has too few resources available.

It is in this step that it is imperative that the founder makes a choice: either they perform the product marketing actions on their own if they have the skills to do so, or they recruit someone else to fill this role. In any case, the role will have to be filled.

1.2.3.6. *From prototype to IPO*

As soon as product marketing has found one or two customers to host a beta test, the iterative product–market fit cycle begins. This test will deliver information that will most likely lead to product modifications, which will interest more customers, etc.

1.2.3.7. *Operational graph*

The entrepreneur started with an idea and after adaptations and changes, he/she made a prototype (see Figure 1.3). The idea is represented by a small lamp, and all changes are represented by an arrow. They resulted in a prototype represented by a V1.

When entering the market, the entrepreneur has expectations. They are represented on the vertical axis. The results can be satisfactory or disappointing.

When the prototype is introduced on the market, a certain level of success is expected. The company then starts looking for a few customers who would be willing to try the product. This test is called the beta test (or proof of concept). In general, in order to get these customers to participate in the test, the product must be modified, all the while it is hoped, of course, that the level of expectations remains the same. Therefore, a certain number of actions will be carried out to modify the prototype that will lead it, with "identical expectations", to a modified product for beta testing (V2). It is the product resulting from the prototype, and modified by customer requests, that will take part in the test.

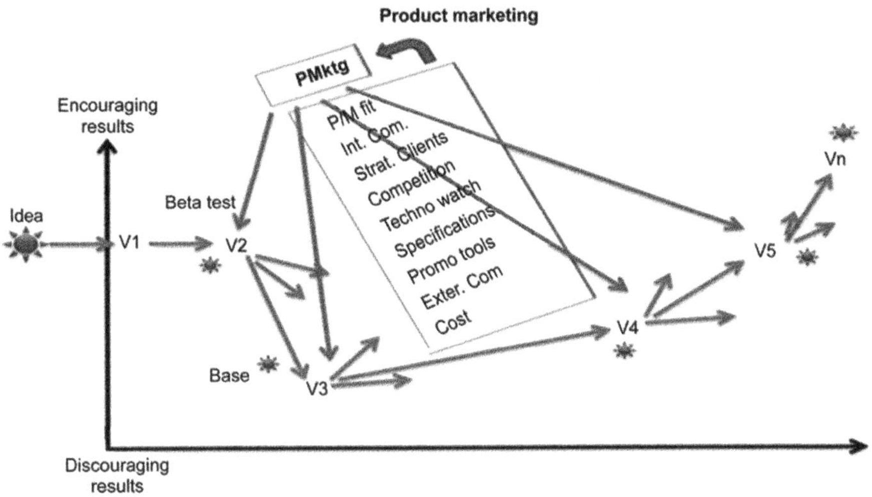

Figure 1.3. *From idea to final product. Product marketing action (source: Georges Kayanakis). For a color version of this figure, see www.iste.co.uk/dispas/innovation.zip*

It is interesting to note that this is the first time the entrepreneur is confronted with the real needs of the market. As we have just seen, this confrontation often leads to product modifications.

During the beta test and at the end of the beta test, it is to be expected that the product will not remain intact. Many changes are imposed by the test results.

In general, the outcome of this process affects expectations. Unfortunately, the new expectations are lower than initially hoped for.

In return for the drop in initial expectations, it should be noted that these actions carried out with real customers – therefore the real market – will lead, by the principle of realism, to a product (V3) that still contains the basic idea, which will have a good chance of being sold on the market.

This important result has been achieved through actions aimed at achieving what is called a product–market fit. It is typical of the product marketing role.

It is important to note that this first product–market fit action may give rise to reluctance, or even resistance, within the company, jeopardizing the chances of success. But in this step, it is inevitable. On the other hand, internal communication work is necessary in order to be able to “sell” the product internally, since it differs from what was initially planned.

It is therefore the first change imposed by the market. It highlights a very important rule: the reality of the market prevails at all times. This is why it is essential that this change takes place as early as possible in the life of the company.

So, after modifications to the previous product and reduced expectations, a product has come to fruition. In principle, it is stable, well-specified, and will be able to satisfy a certain number of customer needs. It can be considered as a basis.

There are now some sound points to develop its sales. It is often that at this point the economic aspect becomes pressing. Obviously, you have to look at the revenue. Often, in this step, the founders have cash flow problems. Having the first customers for a stable product becomes capital. They are often described as strategic, since they help to validate the model and encourage investments.

Finding this type of customer requires specific promotional actions, where the advantage they would have in buying the product must be carefully highlighted (customer benefits).

In this step, the company benefits from a low turnover, rather recurrent, and market knowledge, seen from the inside.

Based on this (V3), the entrepreneur will act with the aim of increasing the revenue and will look for ways to grow for this reason. In order to find more customers, it is necessary, among other things, to:

– redefine eligible markets (target market);

– understand the strategy of the current and future competition;

– carry out technological surveillance;

– impose changes to product specifications;

– develop promotional tools for the commercials;

– carry out external communication actions;

– define sales price objectives (this is the role of product marketing).

Once the previous action plan has been implemented, a target market is determined, which the product seems well adapted for (product–market fit), but which will inevitably undergo slight modifications imposed by the reality on the ground.

Using the same method and at the cost of light and reasonable modifications, we manage to define another product, or another version, which allows the increase of the number of customers and therefore the revenue (V4 and V5).

This step will give the start signal for the economic indicators to improve. Then, we will continue to progress with the same methods, going through steps up to a certain point, Vn, which satisfies expectations.

It should be noted that the initial idea is present in all steps. Here are the two lessons that need to be learned:

– being present and active in the market at an early stage is crucial. Don't wait for perfection to start! We realize that it is the interaction between the product and needs that has allowed us to go from this prototype to a final product. But you still need to know how to carry out these iterations efficiently. This is the role of product marketing;

– having a killer product is an illusion. There are very exceptional cases where a large or small company goes directly to Vn, with a totally satisfactory final product which is then called a killer product.

There are killer products, but it is not often the case. In any case, if a company manages to develop this type of product, it is because it has, from the start, had a very precise prospective vision, in particular:

– a technical and technological vision;

– a competitive vision;

– a vision of the evolution of needs.

Of course, this is possible, but even in this case, it requires at least the existence, from the outset, of a particularly effective product marketing role, which is not often the case. Not to mention, of course, the availability of adequate human and financial resources.

1.3. Categories to represent the progress made

Starting with an idea or a project, forming a start-up to create a profitable, mature company, possibly listed on the stock exchange, requires the completion of some 40 steps (see section 1.2.1).

In the long road ahead, the start-up should, in principle, be enriched at each stage with assets that will be used to take it to the next step, and so on. *These assets are potential actions. Their nature is mainly immaterial.*

These assets allowed it to convert the original idea into a product that sells on the market. This is by definition the innovation process. In other words, meeting the conditions for innovation means owning these assets in time, on time and in quality.

The 40 steps that lead from idea to IPO are divided into six categories that represent the progress of the start-up on this path. These categories reflect the acquisition of certain assets. Each start-up, depending on the path it has taken, belongs to a category.

In order to judge, at a given moment, whether the start-up presents characteristics favorable for innovation, it is necessary to know beforehand, which of the six categories it belongs to.

This categorization can be illustrated with an example:

– referring to the staircase (section 1.2.1), category A, the first by definition, requires the first 11 steps to be completed;

– category B, in the same example, requires the completion of three additional steps that add up to the previous 11. In other words, a company could not belong to category B if it has not passed the steps of category A, and so on. Finally, category F requires that all of the above steps have been completed.

Many problems stem from the fact that some steps that are supposed to have been taken have not actually been taken, thus causing gaps that will subsequently prevent the achievement of certain higher steps.

1.3.1. *Brief description of the categories*

Category A: the start-up has a founder/innovator/creator, etc., alone or with a small team, who has developed a functional lab prototype. There is a vision (present and future market needs), a differentiator(s) for its software product/service and the first simple business plan (1).

Category B: the start-up has a functional prototype that has successfully passed a beta test by one (or more) customer(s). This is the achievement of a fundamental step: belonging to this category proves that the future product has aroused enough interest among customers to accept to perform a test.

Category C: if the positive results of the beta test are confirmed, then a minimum organization is set up to move from the prototype stage to the products, services, software, etc. stage. These are likely to be sold to early customers. An initial fund raising can then be considered.

In category C, there is at least the presence of a technical founder, plus product marketing, or equivalent.

Category D: there is a new business plan (updated business plan 1) taking into account the results of the beta test through the missions of the product marketing: to write the technical specifications of the product, and to update the market needs.

The method of financing is decided and ensured: either by self-financing, in which case the pace of growth must be adapted, or by external financing, in which case growth may be faster.

Category E: there is a virtually complete team: CEO or GM, R&D, sales, product marketing, production. There is a new governance resulting from the funding (see Category D). All the strategies (products, R&D, industrial, etc.) exist. Written procedures exist.

Category F: the team is complete, symbolic and/or significant orders exist. Deliveries and revenue exist. The company “works”.

1.3.2. *Intangible assets to be acquired to move from one category to the next: detailed description and required satisfaction levels*

Moving from one category to the other is the consequence of the acquisition of additional skills that are nothing more than intangible assets. The acquisition, or not, of these skills then becomes criteria.

These assets can also be acquired at different levels (as an example, we can take a student who has acquired mathematics at different levels of satisfaction: low, average, good, very good, excellent).

The following sections summarize, for each of the categories, the subjects that must be mastered by the start-up (which therefore become criteria). For each criterion, the different levels of maturity are described, as well as the tangible elements that the start-up must have in order to ensure that it masters this level of maturity.

1.3.2.1. *Category A*

Vision and/or differentiating product

A vision is what allows you to have confidence in what you undertake, as it often anticipates the future. The present or future vision induces a need to be satisfied, and the question of how to satisfy it: what are the products, services, software, etc. projected?

Often, visions are shared with other visionaries. The difference between two companies with the same vision lies in the differentiation of solutions:

– level 1 is to have a clear, precise, formalized vision and a simple business plan (a few names of potential customers, revenue projections and average sales prices, for example). Tangible element required: a document clearly expressing the vision, and a simple written business plan;

– level 2 is reached when the projected products, services, software, etc., have a differentiator. Tangible element required: the specifications of the projected product/service and a document clearly expressing the differentiators[5] of the product/service and the reasons why they are considered as such;

5 Excluding sales prices – thus costs – unless they are significantly lower than the competition.

– level 3 requires proof that the projected products, services, software, etc., are now being developed by a prototype lab working as planned. Tangible elements required: specifications, lab test reports, photos of the equipment and its dimensions.

Complementary management team

The composition of the management team necessarily evolves according to the company's progress from its creation to the scale-up, and even beyond. During this progression, the nature of the problems encountered, and the solutions they require, requires specific human resources for each step taken. Thus, at the beginning of its life, the company mainly requires technical talents. The management team is reduced to its simplest expression – one person, the founder – and it goes without saying that this person must have these technical skills if the company is a technological one.

As soon as other steps are taken, then other skills become necessary. Although, in general, the founder feels rather competent in everything, the interest of the start-up, which is to optimize its chances of progress, is to have a management team made up of people who are highly competent and complementary in their respective fields. In this way, sterile or blocking overlaps are avoided.

As soon as possible, the manager should relinquish some of their power by delegating it to employees, taking into account, however, that their own profile is part of the equation. For example, if they have a commercial profile, they can delay the recruitment of a sales manager.

Level 1 is acquired when the founder of the project is the technical inventor of the product/service. Tangible element required: the founder's CV.

In case the founder is not the inventor of the planned product/service, then they must imperatively be accompanied, under their direct orders, by someone who is the inventor (inventor included and even if this technical role is largely subcontracted externally). Indeed, this configuration is a better guarantee against the priority dilemmas that arise for the founder. Tangible element required: organization chart and CVs of both people.

Level 2 is reached when a product marketing professional is also present. Tangible element required: providing the organization chart with the job description and profiles of all persons involved.

Protection against competitors (leadership, patents, etc.)

To protect its future, the concerned company must ensure that its projected products/services (regardless of what the competition may be doing) are protected against any threat of copying, counterfeiting or the like.

The first protection consists of marking the date on which the idea, the invention, is made public, thus making any attack impossible due to a patent that would be filed later. To do this, it is sufficient to describe your idea, your invention in a document and to make an official deposit with the competent administration (in France, this procedure is called the Soleau envelope deposit. Most countries have similar procedures).

Note that beyond the relevant intellectual property, start-ups should ideally also have the time and money to enforce it. If this is not possible, it will not become a real competitive advantage.

Level 1 requires proof that the company is in a leading position, by means of a written attestation: a press article, a customer testimonial, a price, an achievement, for example. It is the manifestation of a situation where protection against competitors would come from a product/service advantage over its competitors. Tangible element required: written leadership attestations, documents, publications, etc.

Product marketing organization

Not applicable.

R&D organization and follow-up procedures

In order to better guarantee its capacity to develop its products, services, software, etc., and also to create new ones, a company needs what is usually called research and development (R&D), which is an internal role of a company. As the start-up grows, this role, which is performed by one person (often the founder) at the very beginning, eventually reaches the dimension of an organization, as the task becomes so strategic and complex. At every moment of its life, the company must ensure this role and protect it from other internal considerations that could diminish the effectiveness of this role, which is fundamental to the company's long-term survival.

Level 1 is acquired when the R&D manager is the founder. They have had the basic idea and, as such, are in the best position to develop it; at least in the beginning. Tangible element required when the founder is the inventor (or creator or innovator, etc.): profile and organization chart.

Organized technological and competitive intelligence

In order to not start on a bad commercial track, the concerned company must imperatively have knowledge of what is happening in its market. Indeed, it must check that the projected product, service, software, etc., is not outdated and/or presents *a priori* a more interesting price–performance ratio than the alternatives existing on the market for potential customers.

Level 1 is obtained by the simple fact of "trusting the network", which is, by definition, an unscheduled event. Tangible element required: no document required.

The next level requires access to this type of information through a low-cost promotional action, like a subscription to one or more specialized journals. Tangible element required: written proof of a well-targeted subscription.

Procedure for interacting with subcontractors and suppliers

Not applicable.

Organized risk management

Not applicable.

1.3.2.2. *Category B*

Vision and/or differentiating product

Add level 4 to category A.

Level 4 requires a positive customer beta test demonstrating that the prototype can perform satisfactorily in this customer environment. Tangible element required: the functional specification of the product that has passed the beta test, photo customer test report, with scale.

Complementary management team

Keep the same levels as category A.

Protection against competitors (leadership, patents, etc.)

Add level 2 to category A.

Level 2 requires adding to the existence of leadership attestations, the first step of intellectual protection: the Soleau envelope, which protects against legal action. This package is good protection. Tangible element required: the written leadership attestation and the Soleau envelope.

Product marketing organization

Compared to category A, go directly to level 3.

In order to be able to convert an idea, an invention, a product/service improvement into innovations (i.e. products/services/software, etc., which are sold profitably on the market), it is imperative that a company has a product marketing role specialized in this task.

In the early stages of a company's life, this role is not a necessity, since in these stages, it is mainly a matter of transforming an idea into a prototype lab to verify that the functional objectives are achieved.

Level 3 is acquired when the product marketing role is occupied by a professional. This position, given the modesty of the organization, is not very important, because they are never very far from the founder/manager. Tangible element required: the documents giving the company's organization chart and the Product Marketing job description.

R&D organization and follow-up procedures

Add level 2 to category A.

Whether or not the founder is the inventor of the product/service, level 2 can be achieved when there is an R&D manager present (even, and including, if this is largely outsourced externally). Indeed, this is a better guarantee against the priority dilemmas faced by the founder. Tangible element required: profiles and organization charts.

Organized technological and competitive intelligence

Keep the same levels as category A.

Procedure for interacting with subcontractors and suppliers

Not applicable.

Organized risk management

Not applicable.

1.3.2.3. *Category C*

Vision and/or differentiating product

Keep the same levels as category B.

Complementary management team

Keep the same levels as category B.

Protection against competitors (leadership, patents, etc.)

Add level 3 to category B.

Level 3 clearly marks a threshold crossing. In addition, a patent(s) and a complete analysis of the competition is required. Tangible element required: a copy of the patent filing document plus the written competitive analysis.

Product marketing organization

Add level 4 to category B.

Level 4 can be reached when the person in charge of product marketing reports directly to the business decision-maker (CEO, BU manager, etc.). Tangible element required: documents giving the company organization chart and job description of the PM.

R&D organization and follow-up procedures

Keep the same levels as category B.

Organized technological and competitive intelligence

Add level 3 to category B.

Level 3 requires proof of proactivity, which may be costly, such as participation in exhibitions, shows, conferences, etc. Tangible element required: copies (or proof) of registrations, reports.

Procedure for interacting with subcontractors and suppliers

Not applicable.

Organized risk management

Not applicable.

1.3.2.4. *Category D*

Vision and/or differentiating product

Keep the same levels as category C.

Complementary management team

Keep the same levels as category C.

Protection against competitors (leadership, patents, etc.)

Keep the same levels as category C.

Product marketing organization

Keep the same levels as category C.

R&D organization and follow-up procedures

Keep the same levels as category C.

Organized technological and competitive intelligence

Keep the same levels as category C.

Procedure for interacting with subcontractors and suppliers

Add levels 1 and 2 to category C.

The durability of companies mainly depends on their ability to evolve and innovate. The increasing complexity of technology, regulations, standards, etc., is a handicap in this respect. To consider that a company can be capable, on its own, of grasping most of this knowledge is an illusion.

In addition to the information that can be obtained through the actions of our internal forces (R&D, product marketing, etc.), other sources of information can be used. They are part of the company's ecosystem and share common objectives. They are subcontractors and suppliers and sometimes, public laboratories.

Obviously, this need for collaboration can only be satisfactorily met when the company is mature enough (beyond the start-up), although it appears very early in the life of a company.

Level 1 is achieved by simply having a supplier and/or subcontractor. Tangible element required: the identity of the supplier and/or subcontractor.

Level 2 is reached when there are informal rules of collaboration, such as: visits, meetings to discuss a subject like, for example, a call for tenders. Tangible element required: emails, memos, significant of a collaboration (advice, warnings, suggestions for improvements, etc.).

Organized risk management

Add levels 1 and 2 to category C.

In companies, there are many parameters that can jeopardize its future: technical, financial, commercial, regulatory, normative, political, etc. The edifice can collapse very quickly. Even if the probability of occurrence is considered low, the damage can be fatal. To avoid these dangers, a company must have a role that evaluates the risks taken. As a company gets organized, this role must shift from the responsibility of each functional manager (quality manager, an order at risk by the sales manager, the impact of a customer's bankruptcy by the financial director, etc.) to the responsibility of someone at a high level who is specifically in charge of this subject, thus avoiding the catastrophic pitfall represented by the position of the "judge and judged". A company must be able to evaluate the sum of all risks taken, whatever their nature, at all times.

Level 1 is represented by the situation of normal vigilance that everyone (technical, commercial, financial, social, etc.) in a company is led to exercise in the context of these roles without specific directives. Tangible element required: none.

Level 2 implements limits on various commitments that are believed to pose a potential danger to society. They are generally specified in procedures (e.g. maximum commitment amount for an equipment purchase, commercial guarantees, signature of certain agreements, etc.). Tangible element required: documents (generally, procedures) limiting the commitments.

1.3.2.5. *Category E*

Vision and/or differentiating product

Add level 5 to category D.

Level 5 is reached when a first order of the projected product/service is registered under normal conditions. Tangible element required: a copy of the order and the specification of the product/service concerned.

Complementary management team

Add levels 3 and 4 to category D.

Level 3 is achieved when there is a professional salesperson on the team. Tangible element required: organization chart with job description and profiles of all persons involved.

Level 4 is reached when there is, in addition, a production manager. Tangible element required: organization chart with job description and profiles of all persons involved.

Protection against competitors (leadership, defensive patents, etc.)

Add level 4 to category D.

Level 4 confirms that patent(s) is/are not being challenged and/or that leadership is not being challenged. Tangible element required: authorization for issuance (National IP office or equivalent) and a certificate of non-opposition (National IP office or written statement from the CEO) of the patent(s) and/or clear evidence of leadership (e.g. reports from independent analysts).

Product marketing organization

Add level 5 to category D.

Level 5 requires an independent product marketing organization reporting directly to the business decision-maker (CEO, BU manager, etc.), with sufficient qualified people and an operating budget. Tangible element required: documents elucidating the company's organization chart, job descriptions of the person(s) in the organization and the budget for the current year.

R&D organization and follow-up procedures

Add levels 3 and 4 to category D.

Level 3 requires that the maximum number of ideas, new improvements, etc., are generated. Level 3 requires that the maximum number of internally generated ideas, new developments, etc., aroused, hence the existence and use of study notebooks or equivalent (computer files); the systematic review of the books by the manager ensures that all ideas generated by the R&D organization are transmitted, reviewed and possibly retained, in order to file offensive and defensive patents. Tangible element required: study books (or equipment), past and future planning of reviews, and report of the actions decided upon.

Level 4 must show that the R&D organization is not isolated, but, on the contrary, are open to the outside world and are eager to be at the forefront of knowledge in its field. To do this, staff must have access to journals, articles, newsletters, etc., as well as regularly attend external specialized conferences. Evidence of subscriptions and attendance at external conferences is required.

Organized technological and competitive intelligence

Add level 4 to category D.

Level 4 requires the presence of a professional capable of absorbing the information gathered and using it to improve knowledge of market needs, competitors and their strategies. Tangible element required: organization chart showing the existence of a product marketing professional and his/her job description. Copies (or proof) of registrations and reports. The latest strategic analyses.

Procedure for interacting with subcontractors and suppliers

Add levels 3 and 4 to category D.

Level 3 is reached when collaboration with subcontractors and suppliers is formalized by internal operating rules like, for example, an annual presentation scheduled by the purchasing department, during which the subcontractor and/or supplier presents its activity, skills, products, its strategy. It is a rather passive collaboration, as the content is not directed. Tangible element required: the operating rules (this can be contained in the job description of a buyer, a method engineer, etc.), the identity of the supplier and/or subcontractor and the schedules of past or future meetings, as well as their contents.

Level 4 is reached when collaboration becomes active; the company, depending on its needs, will seek the information it needs from its ecosystem, whether for short-term objectives – like a response to a call for tenders – or for longer-term strategic decisions, like the use of new technologies, for example. Tangible element required: operating rules (this can be contained in the job description of product marketing, or in an internal procedure for CFT (call for tender) approval), the identity of some supplier and/or subcontractor participants and minutes of past meetings.

Organized risk management

Add levels 3 and 4 to category D.

Level 3 requires the means to measure the consequence of each commitment (risk) made, and their possible negative interferences, and to verify that their sum remains safe. These means are represented by a person reporting to the CEO, or the DG specifically in charge of these actions, while adding their critical analysis (e.g. hunting down underestimated risks, etc.). Required: organization chart showing the risk manager, their profile, job description.

Level 4 incorporates the additional notion of transparency in the risks taken. Transparency at the level of the board of directors allows for better measurement and sharing of risks and sometimes for better control or even cancellation of risks. Required: document showing that the risk manager regularly reports to the board of directors on the risks taken by the company.

1.3.2.6. *Category F*

Vision and/or differentiating product

Keep the same levels as category E.

Complementary management team

Add the levels 5 in relation to category E.

Level 5 is achieved when there is a complete organization including quality and finance. Tangible element required: organization chart with job description and profiles of all persons involved.

Protection against competitors (leadership, defensive patents, etc.)

Add level 5 to category E.

Level 5 is achieved when there is a written patent filing policy in place. Required: copy of the patent filing policy.

Product marketing organization

Keep the same levels as category E.

R&D organization and follow-up procedures

Add level 5 to category E.

Level 5 is reached when an adapted training policy is applied. Tangible element required: written policy or three-year history.

Organized technological and competitive intelligence

Add level 5 to category E.

Level 5 is reached when, at least, one product marketing has the mission to do competitive technology intelligence and is given a budget to do so. Tangible element required: job description of the marketing product as well as the budget sheet (or equivalent) for the mission. Also, provide the last two reports.

Procedure for interacting with subcontractors and suppliers

Add level 5 to category E.

Level 5 is reached when, in addition to actively seeking information from its ecosystem, the company will provoke creativity sessions (brainstorming, for example) to better prepare its future. Tangible element required: the last report of one of these sessions.

Organized risk management

Keep the same levels as category E.

1.4. Use of criteria satisfaction levels as a rating guide

The description of the intangible assets to be acquired (and therefore the criteria to be met), and their levels of satisfaction according to the categories, also constitute a rating and/or benchmarking tool.

Indeed, a company belonging to a given category can only move up to the next category if, regardless of whether its financing needs are met, it has the required aptitudes. These are the direct consequence of the acquisition of specific intangible assets (see section 1.3.2), which thus become criteria for fostering innovation and progression towards a possible stock market listing:

1) vision and/or differentiating product;

2) complementary management team;

3) protection against competitors;

4) product marketing organization;

5) R&D organization and follow-up procedures;

6) technological and competitive intelligence;

7) procedure for interacting with subcontractors and suppliers;

8) organized risk management.

Each criterion can be met at several levels ranging from one to five.

2

How to Develop the Innovative Capacity of Companies

2.1. Introduction

In the economic world, it is common to say that a start-up innovates once, when it is born, and that a company innovates as many times as necessary to live, prosper, survive, and to sometimes even be reborn.

Innovation is therefore essential. This goes without saying for a start-up, but for a mature company this implies that it must constantly carry out two activities, which are certainly intertwined, but have different objectives:

– the first activity is dedicated to its social mission as described in its statutes and which monopolizes most of its resources and time;

– the second activity is devoted to the function of "re-creation", i.e. innovation. As such, the assets required are the same as those used for start-ups.

Intangible assets, which can vary from one company to another, are a fundamental source of sustainable value creation (Kaplan and Norton 2004).

In organizational terms, the generation of intangible assets is a holistic approach based on three main and complementary pillars (Yitmen 2011):

– Human capital: "The human capacity of an organization that helps solve business problems" (Bontis 2001). Employees are assets, but they are mobile and cannot be detained by organizations. This category therefore encompasses the efficiency with which an organization uses its human resources (Bontis 2001).

– Internal or structural capital: includes all the non-human knowledge pools of organizations that “stay in the company when employees go home for the night” (Stewart 1997).

– Relational capital: includes relationships with partners (customers, suppliers, etc.) outside the company as well as all other relational resources, such as reputation and loyalty (Sveiby 1997).

Entrepreneurship does not suffer from a deficit of new ideas, but rather from the low survival rate of innovative projects throughout the stages described in the staircase. Indeed, only 20–30% of technology companies survive in the long-term (a good example of one that has survived is the French company SUPRA, which was started in 1873).

Whatever the reason for which a need for innovation arises, facing the question “how can companies develop their capacity for innovation?” is to respond to the challenge of converting these ideas into commercial products that sell for a profit.

In Chapter 1, the staircase describes the conversion process, the following of which improves the rate of survival. The answer to the question is to implement, in a timely manner, all of the actions necessary to move from one step to the next along the staircase.

These actions can be broken down into three parts:

– siphoning queries to produce an idea;

– having the required skills;

– having a focused organization.

2.2. Siphoning queries to produce an idea

The first action to undertake, in order to increase the capacity for innovation, is to be able to channel, or siphon, all these requests for innovation towards decision-makers, through routine actions that can give them visibility. We mention a few effective measures: permanent maintenance of R&D notebooks (digital or otherwise), monthly review by management:

– R&D notebooks;

– “innovation” requests from the product marketing and product engineering departments.

The objective is to produce an idea, a differentiator, for a start-up or a company that allows one to either maintain and improve its position, or to address new needs and/or new opportunities on the markets in line (or that are at least similar) with the general makeup of the start-up or the company.

2.3. Having the means (skills) required

Generating and selecting ideas is, of course, not enough. The company, whatever its size, must acquire the skills to:

– technically succeed;

– protect against dangers;

– generate routines and procedures.

2.3.1. *Technical achievements*

Starting from an idea, the company must have the skills to carry out a certain number of actions strictly necessary for it to have a chance to technically lead to an innovation:

– make a prototype with the means at hand (and keep it inexpensive) for a start-up, and R&D for a company;

– write technical specifications of the product and define cost objectives (perform a positive beta test), identify corrective actions;

– carry out corrective actions;

– write the new modified technical specifications;

– check the product/service–market fit;

– enter the first symbolic and then significant orders.

The nature of these skills mainly involves know-how from human resources:

– for the start-up: the founder, an engineer (if the founder themselves is not an engineer), product marketing;

– for a mature company: a product manager, product marketing, an R&D manager.

It should be noted that at no time is the sales organization mentioned. In fact, the sales organization is hardly involved in the innovation process, as it can only be effective if it has stable, well-specified products/services available on specified dates and at a precise price. However, it participates indirectly in this process by giving product marketing access to the market through appointments, presentations, advertisements, etc.

2.3.2. *Protection against dangers*

During the first period of the start-up's existence, founders tend to devote themselves exclusively to the concrete implementation of their ideas. Making the prototype then becomes the main goal of their activities. However, many months (or sometimes years) might pass between the idea being born and the emergence of a number of dangers: for example, alternative solutions, new competitors, unexpected costs, unfavorable regulations. It becomes urgent to build the following two defenses as much as possible:

1) An increase in the protection against competition (Soleau envelope, model, brand, patent, etc.), as reported in the literature (e.g. Halt *et al.* 2017). Start-ups need to think about trademarks, patents, designs, copyrights, intellectual property, labeling and marketing issues and, of course, nowadays, domain names. Failure to protect and enforce one's intellectual property can sometimes result in a decision to not proceed with the development process. However, beyond this "defensive" aspect, intellectual property can be seen as a real lever to create and pursue market opportunities in existing or new markets (Conley *et al.* 2013). It should also be noted that intellectual property will be valued when potential fundraising emerges. Subject to the availability of the necessary financial means, this protection can also be obtained through constant leadership. These protections call for the following HR: R&D, product marketing, founder or manager, risk

manager, and lawyer. As these last two roles are not full-time (at least in the start-up phase), they can be partly filled by other people or outsourced. For example, at the very beginning, the company may call upon a specialized lawyer/intellectual property attorney, or the risk management role can be performed by the founder, and then by another person (e.g. a quality manager).

2) Managing risks: the company and its ecosystem evolve in environments where risks of all kinds exist. Managing risks means being aware of them in order to avoid and anticipate them (see section 1.3.2).

2.3.3. *Generating routines and procedures*

Routines and procedures are systematic actions, and rules, whose purpose is to ensure that the company is not subject to chance or, at best, the random talents of a few. Here are the main routines and procedures that must be established:

– monitoring the competition (permanent);

– monitoring product/service–market fit (permanent);

– R&D procedures: how to generate and siphon ideas, solutions, etc. (permanent);

– having technological surveillance (permanent);

– working with the ecosystem to have more intelligence, information and solutions (permanent).

Implementing routines, procedures, etc. requires a willingness on the part of the founder (or the GM) to impose them, and on the part of product marketing, R&D and the production manager to execute them.

2.4. Having an appropriate organization

2.4.1. *Prerequisites*

Regardless of the category the start-up belongs to, an innovation process cannot take place without coordination, decision-making and an organization chart, however simple it may be. This organization must be such that at no time can there be a pause on the priority: innovation.

An example of a common dangerous situation in a start-up is that the major product marketing role that needs to be performed by a professional is sometimes either under the orders of the sales director or combined with the roles of the founder or GM. This type of organizational structure does not reflect the priority that must be given to innovation. In most cases, it is strongly recommended that the person in this role reports directly to the founder or MD. In young start-ups, if the founders are themselves trained in product marketing, they can, for a certain period of time, ensure that this cumulation runs smoothly.

2.4.2. *Building the capacity to innovate*

The means described above (see sections 2.2 and 2.3) are used to create skills. The latter are, in essence, part of the company's assets.

Answering the question "how does one develop the capacity to innovate?" is the same as answering the question "how does one acquire the intangible assets that enable us to move from an idea or an invention, to a product/service that sells profitably?"

The ability to innovate is therefore reduced to the acquisition of the intangible assets necessary to fully carry out the innovation process, as defined above. These assets generate skills, the presence or absence of which are criteria for judging the ability of the start-up – or the company – to innovate.

2.4.2.1. *How and when should these intangible assets be put in place?*

As we have seen, this process is the same for a start-up or a mature company looking to innovate in order to achieve important objectives, such as reducing costs, developing new products and amplifying human action.

The implementation of these intangible assets, therefore of skills, can be done in the following way:

– *Naturally*: when the start-up or company considers that establishing and maintaining these skills is one of its objectives. The main condition is to be aware of their existence and importance.

– *Deliberately*: following a stagnation of the start-up or company that has highlighted problems to be solved.

– *Occasionally*: when seeking external funding, gaps appear and require action in order to be filled.

2.4.2.2. *Intangible assets needed: criteria for identifying them*

For this purpose, a few criteria (benchmarks) are defined to be fully representative of these assets. They result from a reading and understanding of the staircase as a chronological sequence of events leading to success, completed by an operational reading of this same staircase, which stipulates the transversal assets (or intangible skills) outside of the chronology. For example, having an idea and differentiation is an asset that must be present at all stages of the staircase, whereas having a prototype is an asset that is no longer necessary once the products delivered are satisfactory (see step 18 of the staircase: business plan 2).

2.4.2.3. *List of criteria*

It consists of the following elements:

– a vision, a product and one (or more) differentiator(s);

– complementary management team;

– protection against competition;

– product marketing organization (product–market fit);

– R&D organization (or capacity to innovate) and operating procedures;

– technological and competitive intelligence;

– procedures for interactions with subcontractors and suppliers;

– appropriate risk management.

2.4.2.4. *Content of the criteria*

They are defined and explained in Chapter 1.

2.4.2.5. *Functioning of the criteria*

2.4.2.5.1. Need for gradation of levels in each criterion (Diag. O method)

While certain assets (skills) are essential to complete the successive stages, they must also be satisfied at the right level. Hence the gradation of these: level 1, level 2, etc.

Thus, a student, in order to move up to the next science class, must meet the following requirements related to the criterion "being good at math". Based on the A, B, C, D, E and F level of the student's grades, it is easy to infer that their acquired assets (abilities) in math predict that the higher their grades, the better their chances of moving up to the next grade.

On the road to success, and in order for a company to become more mature, it should, in principle, meet more and more criteria, and meet them more and more fully, as it progresses through the A, B, C, D, E, and F categories.

These criteria are simple, but often forgotten along the way. They must be met at all times, regardless of which stage of development the company is at.

Let us examine an example of how the criteria work, and the different levels of satisfaction, through the following example: "having a vision, a product and one (or more) differentiator(s)". This criterion is essential to guide the company's actions; it is a foundation the company is built on and must exist at all times of the company's life.

"Having a vision", the first level of satisfaction of this criterion, has a structuring impact.

This vision has induced a mission (*raison d'être*) that has, as we have seen, become in turn a criterion, since its constant presence is indispensable. As the company is structured from the outset according to this, constantly satisfying this criterion gives the company the ability to be driven without the risk of being diverted and without dispersing its efforts. Moreover, it induces the company's personnel to work cohesively and also allows the company to be perfectly identifiable in the market.

The banal phrase "everyone works here towards the same goal!" is testimony to the value attached to a common and shared vision within the same company. It inevitably implies that the company is particularly well-equipped to face what lies ahead.

"Having a vision, a product and one (or more) differentiator(s)" is the essential minimum needed in order to satisfy this criterion. Having a differentiator is the essential asset that must be met to create a true competitive advantage. This criterion is linked to the vision criterion in the following way: a vision can be shared by two competitors, but it is necessary to have a product differentiator in order to distinguish oneself. This is a very important criterion, since it leads, in particular, to shaping R&D efforts, a selection of suppliers and a precise focus of advertising and marketing efforts. Thus, it makes it possible to answer the essential question: "Why would a customer come to me rather than to a competitor?"

As this question arises throughout its life, the company must constantly meet this criterion: to have a differentiator, whatever it is (provided it is valid, of course), and provided that the company is aware that any modification not perfectly controlled leads to a dangerous zone.

When a modification is envisaged, it is necessary to look at all the figures of the company again, re-analyze the competition and possibly look at the positions seen above again: R&D, suppliers, etc.

2.4.2.5.2. Required criteria levels according to categories (Diag. O method)

The levels required to meet a criterion vary from category to category.

To illustrate this, we can take the example of the first criterion, which is "having a vision, a product and one (or more) differentiator(s)"; this criterion would be met by the category A start-up if it had a vision and a prototype, whereas for a category D start-up, the following would have to be added: a positive beta test, plus a first order, etc.

EXAMPLE.– Application of the criteria: difference in requirements, for the same criterion, between two categories (A and D).

To illustrate this example, let us recall the criterion as defined in Chapter 1 (see section 1.3.2): a vision is what gives you confidence in what you are undertaking, as it often anticipates the future. The present or future vision induces a need to be satisfied, and then how to satisfy it: what products, services, software are projected?

Often, visions are shared with other visionaries. The difference between two companies with the same vision lies in the products/services and the differentiation they present.

Let us explain the levels of satisfaction required according to the category that the company belongs to. During its progression from category A to F, the company must have a vision that progresses from a simple statement that fits on a piece of paper, to its complete integration in both its daily and strategic actions. This progression of the vision can be broken down into levels; the passage from a lower level to a higher level is achieved by the integration of an additional skill.

In this case, here are the different progressive levels of satisfaction that must be achieved:

– Level 1: the minimum is to have a clear, precise, formalized vision, and a simple business plan (a few names of potential customers, sales projections and average sales prices, for example).

– Level 2: is reached if the projected products/services/software has a differentiator.

– Level 3: is reached if the projected products/services/software has now come to fruition by means of a lab prototype that works as imagined.

– Level 4: requires a positive beta test from a customer, demonstrating that the prototype is satisfactory in a "market" environment.

– Level 5: is reached when a first order of the projected products/services/software is registered under normal conditions.

Let us apply this to category A: in this category, it is normal for the company to have a prototype in its lab, however rudimentary it may be, that is representative of its vision. In this case, the level reached is 3.

Then let us apply this to category D: on top of what is required for A, we have to add proof that a positive beta test has been performed at a client's site. In this case, the level reached is 4.

3

Innovation and Its Challenges for the Company

3.1. Introduction

Innovation is a malleable concept: it is often distorted according to need. Let us not confuse creativity, inventiveness and innovation:

– creativity is a set of techniques for generating ideas which can be acquired;

– inventiveness is the ability to produce inventions, often an individual talent;

– innovation is not just the preserve of start-ups, but an invention that is sold or exploited profitably and may not necessarily be technical;

– innovation seems to have variable content: content varies according to the need for action. It has long been the focus of attention for economists and other leading entities. To understand and, ultimately, to master it, many have tried to determine the content. For example:

- Schumpeter: innovation can concern the process, the organization, the product, new sources of raw materials and new markets. He stressed that innovation must be distinguished from invention. The reason Schumpeter emphasized this difference is because he considered innovation to be a specific social activity, or "function", carried out in the economic sphere and with a commercial purpose, whereas, in principle, inventions can be made anywhere and without any intention of commercialization. Thus, for Schumpeter, innovations are new combinations of knowledge, resources, etc., which are subject to attempts at commercialization – essentially the process by which new ideas are generated and put into commercial practice. He called this "combinatorial" activity the "entrepreneurial function" and the social agents who perform this function, the "entrepreneurs",

- the Oslo Manual (OECD): innovation concerns the product, the process, the marketing and the organization,

- Christensen: who speaks of innovations of continuity, efficiency or transformation.

Innovation applied to the company addresses the essential: *its survival, its growth, its profitability*.

All of these definitions end up being relevant depending on the company concerned or the moment in time. They only serve to break down the concept to make it more manageable.

The most important thing in innovation is to know what can be expected from it: wealth creation, growth and survival.

In fact, the inception of a company does not guarantee its survival, growth or prosperity; it is essential that it possess the necessary tools to face adversity and to solve any difficulties to achieve these objectives.

Innovation is the best of these tools. Under the general umbrella of the company's "innovation policy", it can be broken down into several categories that address *major sectoral issues*:

– improve and create new products/services;

– *create differentiation, a strategic advantage*;

– reduce production costs (processes, raw materials);

– *establish oneself as a leader*;

– create new markets (digital marketing, etc.);

– create a competitive advantage based on regulations (CSR, GDPR).

3.2. General innovation policy: a vital issue, survival

Innovation is the best way to meet the challenge of survival, growth, and corporate enrichment. *But first of all, survival!* This assertion implies that *the company must constantly recreate itself while maintaining the imperative: on time.*

In this context, recreating oneself is to engage in a special process whose benefits are not immediately visible, and which may conflict with more short-term interests, especially financial interests.

An econometric study conducted on 18,243 French companies with more than 10 employees also confirms that innovation and growth are interlinked (Levratto *et al.* 2011). Indeed, this study concluded that "innovation, acting in concert with a set of structural (group organization) and environmental (sector) factors, positively influences revenue growth. Its effect is clearly more noticeable for firms experiencing very strong growth in market share than for others. On the other hand, performance is not very sensitive to the type of innovation chosen."

3.2.1. *Uncertainties and costs of innovation, in return for survival*

Embarking on a process of innovation means entering a period of uncertainty: commercial, financial, human, organizational, etc.

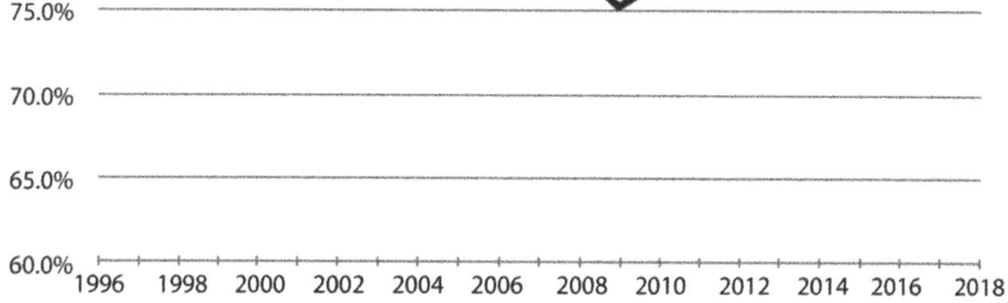

Figure 3.1. *Survival rate of start-ups after one year of activity*[1]

1 Source: Kauffman indicators of Entrepreneurship (2018). National report on early-stage entrepreneurship.

Moreover, the survival rates of start-ups illustrate the difficulties that every innovator faces. In their first year, 20% of American start-ups disappear.

After five years of activity, about one out of two companies has officially disappeared. This is not only the case in France, where the 5-year survival rate was 52.6% in 2000 – according to DATAR (French regional land development agency) and INSEE (French national institute for statistical and economic studies) – but is also true in the United States, as can be seen in Figure 3.2.

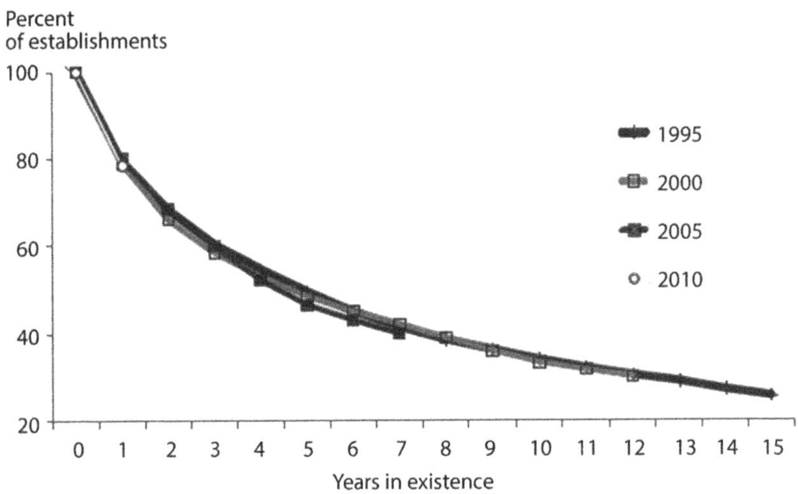

Figure 3.2. *Survival rate of business start-ups by year of establishment*[2]*. For a color version of this figure, see www.iste.co.uk/dispas/innovation.zip*

However, for a tech start-up, survival is even more difficult than that in reality:

– official statistics take into account so-called "zombie" companies, i.e. companies that are still registered but no longer have any operational activity;

– as illustrated above, mortality continues to increase after five years;

2 Source: Bureau of Labor Statistics (2014). Small Business Administration, Office of Advocacy, FAQ.

– finally, these survival rates concern all sectors of activity. Mortality is even higher in the technology sectors. Indeed, a professor at Harvard Business School (Gosh 2012) studied more than 2,000 start-ups that raised more than a million dollars from venture capitalists and concluded that three-quarters failed. This order of magnitude is corroborated by a study by CB Insights (2018), which analyzed more than 1,000 start-ups that raised funds and found that 67% of them either filed for bankruptcy or barely survived.

CB Insights has also analyzed the top 20 reasons for the failure of more than 100 start-ups that have raised funds.

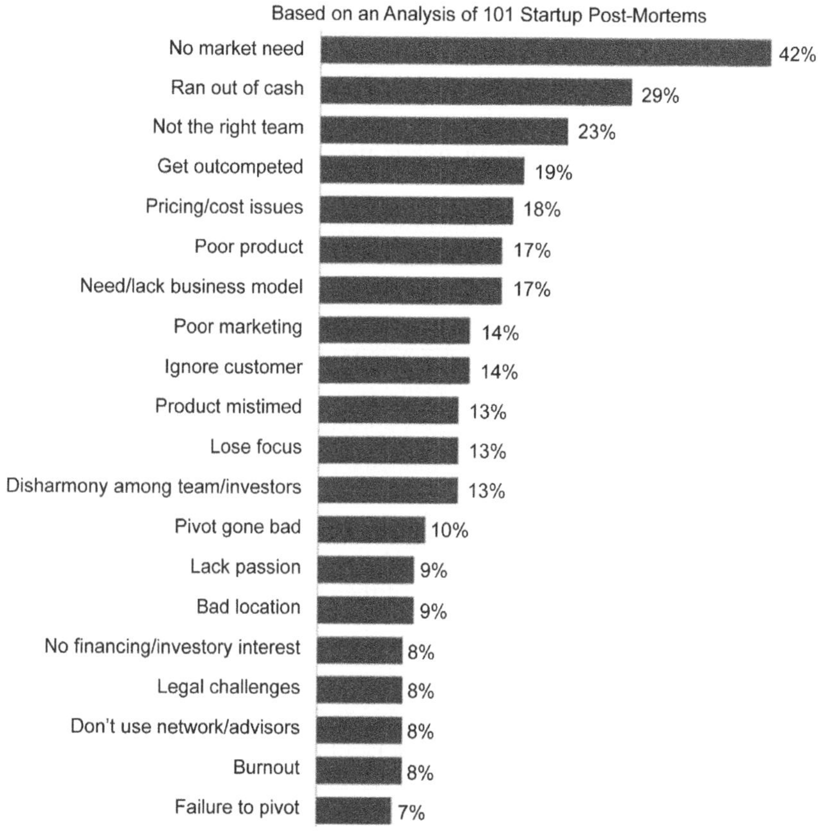

Figure 3.3. *Top 20 causes of start-up failure*[3]. *For a color version of this figure, see www.iste.co.uk/dispas/innovation.zip*

3 Source: CB Insights (2019). The Top 20 Reasons Start-ups Fail. Report.

What kind of conclusion can we draw from these rates, knowing that a mature company must innovate *n* times throughout its life?

The difference between a start-up and a mature company is that the former, unlike the latter, leaves without any assets and often lacks financial means.

The fact that it exists proves that a mature company has been able to innovate at a given time. Consequently, it has had the necessary assets to do so. Has it retained them? This is a question the company must answer. If necessary, and if aware of it, the company must make up the shortfalls; otherwise, it will fail in the long-run. These gaps can be difficult to fill. How many of the world's market leaders, at a given time, have missed certain turns of events that have led to colossal losses of market share, or even halted their activity? Examples include Nokia or Blackberry in phones, or the search engines Lycos, AltaVista, Excite or Ask in the Internet search sector.

Innovation is no longer the business of a single person, or of a few people, knowing that (to take the example of R&D only) soon the most brilliant employees, alone and isolated in the company, will no longer be as worthy as an external collaborative network, connecting the best in the field concerned. It is clear that innovation will no longer depend on the presence of an exceptional person within the company's core business as much as it did before. It will depend much more on the – uncertain – ability of the collaborator to integrate into networks and to feed off what is happening elsewhere.

In other company roles, the challenge will be to invest in real assets (organization, procedures, HR), to create the right conditions for innovation without disrupting the smooth running of the company.

3.2.2. *A deadly risk*

Schumpeter said (Schmitt and Cohen Boulakia 2019):

> As with humans, companies are born without the strength to survive. Others succumb to what one would call, for a living being, a death "by accident" or "from the consequences of an

> illness". Some others disappear from a "natural" death, just as we die of old age. This "natural" cause, in the case of companies, is precisely their inability to keep up with the pace of innovation that they have imposed on themselves in their good times.

To continue where the above left off, it can be said that an appropriate pace of innovation can prevent companies from succumbing.

That said:

– a company in a period of stagnation often thinks that embarking on a long and costly innovation operation can put it in a risk zone and later in a dead zone. In reality, returning to the discomfort zone, where the company found itself some time ago, corresponds to a feeling of regression (often a reality) that is difficult to accept in the short-term, but often leads to salvation in the medium-term;

– a growing company often thinks about stabilizing its performance and/or increasing its revenue by taking so-called "controlled" risks.

What do the words "stabilize" and "controlled risk" really mean in our times? Who really wants to take uncontrolled risks? All this really means is that resource trade-offs will be made in the spirit of a compromise: of course, a bird in the hand is worth two in the bush. But as Schumpeter said, we must keep up! He defined what was worth two in the bush as follows: it is what will make it possible not to succumb in the shorter or longer term.

3.3. Innovation: major sectoral issues

If company management has decided to invest in innovation, then an analysis must be made to specify the sector(s) to be addressed.

Depending on the diagnosis, it may be necessary to concatenate several sectoral actions (e.g. innovating to create a new product/service may precede commercial innovation through the use of digital marketing).

3.3.1. *Innovation? Improving and creating new products/services*

Creating and introducing (development and introduction) new products/services take time. Figure 3.4 shows that as the useful trading period (growth

+ maturity) becomes shorter (the decline occurs faster than a few years ago), it is imperative that the replacement product be developed at the end of the introduction.

This is all the more problematic because even the development and introduction period is shortening due to the multiplication of competitors or competing solutions acting in a competitive market: price wars based on cost reductions; erratic and competitive appearances of new techniques/ technologies; acceleration of the trend to integrate more and more features, etc. A shorter development and introduction period that requires a safer and more efficient process induces the strengthening of innovation capacities in order not to be overwhelmed. This is all the more true as new (and/or replacement) markets appear very quickly, accompanied by a procession of competitors, unknown the day before.

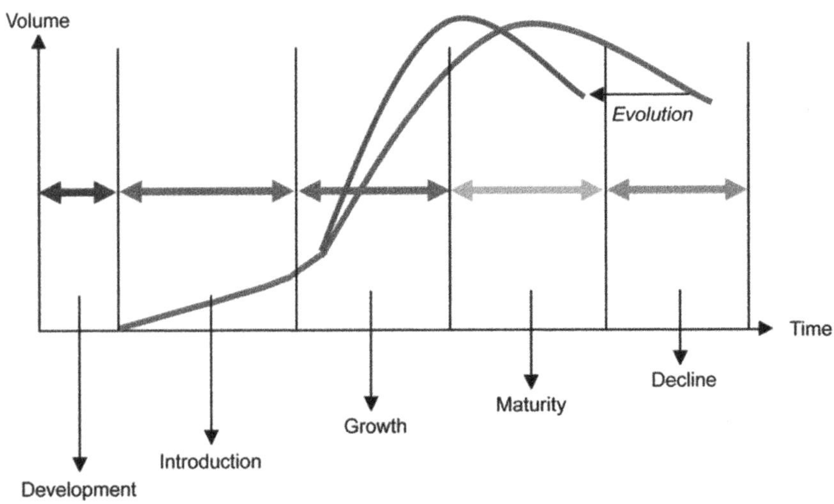

Figure 3.4. *Life cycle of a product/service. For a color version of this figure, see www.iste.co.uk/dispas/innovation.zip*

Faced with these challenges, the company is obliged to respond with a new product/service offering. This response cannot be produced overnight.

In all cases, a quick response is required. This will be all the more important as it will have been prepared in advance, continuously (often spread over several years) and in offensive mode, thanks to technological intelligence, monitoring of competitors, customer contacts, etc.

In such a situation, it is necessary to focus on the areas where the company has assets:

– eventually redefining the scope of its market (perhaps refine it) and, if it is new, segmenting it (target market versus TAM);

– redefining its objective status (leader, follower) and the associated metrics;

– regularly redefining its strategy and plan to achieve this.

3.3.2. *The challenge of differentiation*

Differentiation is what allows a company to create a significant advantage (benefits) for its customers, more valuable for the target audience than that of its competitors.

Having a product/service with a differentiator is a major strategic advantage, since it puts the company in a preferential position compared to its competitors.

This advantage is most often created through innovation, the goal of which is either to improve or create a product/service. It is obvious that such a strategic advantage cannot be defined without in-depth knowledge of market needs, their possible evolution and the (current and future) products/services of the competition.

Several types of competitive advantage exist. As a non-exhaustive list, we can mention the following:

– *a cost advantage*, therefore a margin advantage for the *customer*: reduce the start-up's sales prices for its customers, which favors the margins of the latter;

– *a reliable quality advantage*:

- quality is an effective short-term advantage: the short time between delivery and first use, the rate of non-conforming products, and defective systems is a real discriminating factor,

- reliability is a great advantage in the medium- to long-term. The risks of breakdowns/catastrophic failures with disastrous consequences, if they occur during use, should be avoided. The supplier with the lowest failure rate – or the highest MTBF[4] – has a real and powerful differentiator;

– an advantage to facilitate the customer's product/system road map: the company's customers also need to evolve their products/services. The supplier that anticipates this creates a significant differentiator, since they help their customers to progress;

– an implementation and simplification advantage, for the customer: the manufacturing cost of a product or system includes the cost of material (or equipment), labor and expenses. When a product/system is studied in the customer's interest, the advantage of easy integration in manufacturing (low waste) is complemented by a labor minimum. This advantage improves the customer's gross margin and favors the supplier.

An example of differentiation: Vite Mon Marché.

Traditionally, and particularly since the advent of mass distribution, the value chain of agricultural products includes many actors and links that share the added value resulting from the difference between the price paid by the end consumer and the cost of production of the producers/farmers.

According to the French Ministry of Agriculture and the French Ministry of Finance, when considering domestic food consumption, agriculture recovers only 10% of the total induced value added. The rest is recovered by the various intermediaries (18% by the food industries, 24% by inter-firm trade, 22% by services, 21% by catering and the rest by other industries). This is also the case in the organic products segment, 49% of which are distributed by mass distribution; 34% by specialized distribution and 5% by artisans, traders and restaurants. This means 88% of organic production is sold via intermediaries who each take a margin for the service they provide to end consumers, against only 12% for direct sales (Agence Bio 2018).

At the same time, several major trends are emerging in society:

– the explosion in demand for organic products, which increased from 1 billion to 10 billion euros in sales between 1999 and 2018 (FCD – *Fédération du commerce et de la distribution* 2019);

4 MTBF: Mean Time Between Failures.

– the growth of home delivery in all sectors of the economy as illustrated by the growth of e-commerce giants (e.g. Amazon) or the explosion of food delivery services (Uber Eats, Deliveroo and Frichti);

– the uberization of intermediaries with, for example, the exponential growth of all platforms directly linking a creator/producer with an end consumer (e.g. Uber or Canva) or cryptocurrency (which is exchanged from peer to peer without institutional intermediaries);

– the rise of conscious consumers who choose to consume in short circuits in order to make their local ecosystems work and reduce the carbon footprint of their consumption.

Based on these various observations, the company Vite Mon Marché, initially launched as a marketplace for farmers, has innovated operationally and logistically to build a unique value proposition: "Vite Mon Marché is the direct market of farmers delivered to your home in less than 60 minutes at a rate similar to that of specialized distribution."

Consumers can choose from a wide range of products that are 80% local and 65% certified organic. In addition, the local producer also benefits from a reduction in the number of intermediaries, since he or she recovers more than 50% of the final selling price of their products.

While many similar initiatives exist, the differentiation of Vite Mon Marché comes from the fact that this company simplifies the life of the consumer who can get a delivery within the hour. This is made possible, notably, thanks to logistical and operational innovation. Indeed, for example, in the city of Nantes, seven trucks are permanently optimized to pick up products from local producers and deliver within the hour to geolocalized customers who have ordered via their app.

And this operational innovation, which corresponds to a real customer need, is working: barely a year after its launch, Vite Mon Marché prepares and delivers more than 800 orders per week within the hour and generates a revenue of €1.5 million. The company raised €2.3 million in June 2019.

3.3.3. *The challenge of being the leader*

There are two main types of leadership (they are not mutually exclusive):

– market leadership;

– technological leadership.

Being the market leader means the company's turnover, or the volume delivered – sometimes both – exceeds that of any other competitor.

This winner-takes-all strategy can be illustrated by Google, Amazon and Facebook, which in the space of a few decades have managed to climb the ranks to become world-leading stock market capitalizations and gain market share as oligopolies.

Indeed, depending on the different cases of competition (monopoly, oligopoly, atomicity), the innovation intended to conquer new positions, or to confirm others, takes on different aspects conditional on the possible stakes (maximization of profits or penetration, etc.):

– for a monopoly, with a maximum market share and turnover, only defensive innovation is necessary, to maintain prices and ensure an excellent quality, reliability and service trio;

– for an oligopoly, market shares are no longer disputed and competitors charge very similar prices. The danger of losing our share comes from the impossibility of rapidly lowering sales prices and therefore production costs. Consequently, the innovation of rationalization (cost reduction) will be mandatory in a preventive perspective (having a cheaper version in reserve);

– for a fragmented market where it is particularly difficult to have a differentiator, where market shares vary according to open competition, it is necessary, if the market is attractive, to develop a replacement innovation (new product).

Being the technological leader means the company's strategy is based on the "brand image" effect. This is acquired through recognized expertise and its corollary: reputation. The development of "audacious" products or those well ahead of their time and the associated communication are sine qua non conditions to achieve this. Having a technological leader as a supplier encourages those customers, who are the most numerous, who either want to take care of their own image or not take risks.

The position of technological leader is very favorable, because the customer's personnel involved in the choice of supplies most often want to work with leaders in order to benefit from some kind of guarantee. In this case, the criterion of "leader" joins that of "differentiator".

3.3.4. *The challenge of reducing production costs*

Reducing costs is a fundamental duty of the company. Indeed, under the pressure of global competition, the prices of products/services are steadily falling, reducing margins accordingly.

Faced with a sharp drop in market prices, the company is faced with an alternative: lose market share by maintaining its prices or lower prices and reduce its margin. In both cases, the company loses out.

The normal response to declining market prices for a product/service is to reduce costs along with selling prices to maintain (or increase) margins. However, given the time required to reduce costs, it is fundamentally important that this action be permanent and systematically included in the business plan/budget.

With an identical product/service (same technical specification), reducing costs requires innovation in at least three areas, those of their components:

– the purchase price;

– the direct cost of production labor;

– the indirect cost of production expenses (management, quality, fluids, electricity, space, etc.).

Depending on whether it is a product or a service, these components have a different weight. For example, a pure service like accounting outsourcing mostly has components 2 and 3, while a pure product like a cell phone has all three components.

– Reducing the purchase price of raw materials, components, materials: these purchases represent direct costs, and reducing them is beneficial to any start-up/company. This reduction does not require innovation efforts, except in the case of a specific production requested from the supplier (custom design) that will require some counterpart, like a higher price, for example. In this case, the R&D resources of the start-up/company are used for studies that replace those of the supplier. Most often, it is decided to proceed with "standard" purchases: materials, components, equipment, etc., at lower prices from suppliers. In this case, there is nothing strategic about reducing the costs resulting from this purchase, since all competitors can benefit from it.

– Reducing direct labor costs simply means that the time taken to manufacture is reduced by an innovation and more than offsets normal wage increases. The time taken to manufacture or produce (a product, software, a service) depends on a few important parameters: the productivity of the equipment, its ergonomics, breakdown rates, yields (reject rates), organization (or process), lost time, etc., and then the training and motivation of personnel. Changing production means taking a risk; the long list of parameters being important and the time for action to bear fruit being very long, prohibits frequent changes in manufacturing conditions. Any change must be prepared in advance, including the transition. In order to reduce the direct manufacturing time, innovation will be of interest to a wide range of fields such as new equipment, new software, new processes, new materials, and new components and hardware.

– This action should be routine, but at a low frequency for its heavy parts, like equipment, and at a higher frequency for their availability rates. Whether for materials, components or equipment, or for labor costs, the cost reduction function offers the company the additional benefit of keeping in touch with technological developments practiced by suppliers and subcontractors, thus reinforcing its image as a leader.

– Reducing the indirect cost of production expenses. It is an indirect cost, and therefore fixed or pseudo-fixed. Since it varies with volume (it decreases when volumes increase and vice versa), it may not seem to be a priority in periods of growth; on the other hand, wisdom would dictate that it should also be subject to innovation when volumes are no longer there. In order to reduce these costs, it is necessary to call upon organizational innovation (Oslo Manual) in a specific context: a program to align the direct/indirect ratio with comparable organizations.

3.3.5. *The challenge of penetrating new markets*

The company, faced with the need to naturally progress, turns to what is commonly called its "growth drivers".

Based on the company's DNA, ancillary or related markets have, in principle, been identified for a long time. It is then natural to try to conquer them, because the "jump" (gap to close) to be made seems limited.

The advantage of entering a "growth relay" kind of market is considerable from the point of view of financial performance: the marginal effect on a certain number of company roles such as sales, R&D, HR and control/finance departments and purchasing produces excellent results.

In other words, "growth relay" means proximity, but not identical.

This difference has to be made up by a complete innovation of a new product/service that will have to be carried out by going through all the steps of the innovation process, except the first ones, since the idea is already acquired. It will be necessary to create a differentiator, make a prototype, carry out a *beta test,* check the fit with market needs, enter the first symbolic and then significant orders, etc., and, finally, to have the funding to make progress.

Innovation may be sought in the area of advertising and marketing. Even in this field, proximity does not mean identity.

Of course, access to these growth drivers is a performance that can in itself justify an IPO. Mastery of a precise DNA, proximity of markets, and past performance give confidence in financial forecasts.

3.3.6. *The challenge of creating a competitive advantage CSR and/or GDPR*

3.3.6.1. *CSR (Corporate Social Responsibility)*

Putting a product/service on the market today has important societal consequences. Companies, their environmental footprint, their products and their waste are considered threats to the planet. The appropriation by all of this vision has an impact on the choice of suppliers, because:

– citizens are breaking into companies;

– planet Earth enters the company.

Not so long ago, competition was considered the engine of the economy and social welfare; innovation was the fuel for it (Chicago School – Milton Friedman – Nobel Prize). Since then, many things have changed.

With the arrival of countries accompanied by their procession of new competitors in the economic world, the pressure on prices has become such that outsourcing has multiplied at a crazy pace, leaving aside all the social consequences.

But the financialization of the economy has begun to lose its followers since the 2007 crisis.

The value of intangible assets (especially HR) is gradually regaining its rightful place. At the same time, the inhabitants of the Earth are becoming increasingly aware that the permanent and tireless search for lower costs, under no less permanent competitive pressure, could lead to catastrophes on a global scale.

Faced with this rising collective awareness, and as a first step, the company responded by adopting an avoidance strategy: avoiding conflicts with stakeholders, avoiding legal risks and, above all, avoiding banishment. Then, faced with the persistence of this generalized reproving awareness, in a second stage, it adopted a strategy of accompanying and then anticipation.

There are four main justifications for implementing a corporate social responsibility policy (Porter and Kramer 2006):

– moral obligation (respect duties, respect citizens, respect the environment, behave ethically);

– sustainability (meeting the needs of today without affecting future generations);

– work permit;

– reputation (brand image).

They explain why the first two justifications are intimately linked to strictly immaterial factors related to female/male citizens. The third justification is strictly technical, while the last one clearly states that a "business" benefit is sought through the first three points.

These same authors also consider that the better development of human capital constitutes new strategic opportunities for differentiation and the constitution of a competitive advantage.

Porter and Kramer's four justifications gradually became real arguments in favor of the company, its sustainability, growth and profitability.

The objective of CSR innovation is to build a structured competitive advantage with this set of intangible assets (except for the third). Knowing that in these matters there is interdependence between the company and the territories in which it operates, CSR innovation addresses very broad fields.

3.3.6.2. *GDPR (General Data Protection Regulation)*

European citizens benefit from new rights on their personal data, such as portability, dereferencing on search engines and oblivion.

The obligations created for companies are both considerable and strictly necessary for their survival, since failure to comply would jeopardize the recruitment and compliance that will be unfailingly demanded during calls for tenders.

These new obligations must also be consistent with other important considerations, for example:

– as personal data has often become the fuel of the digital world, this regulation should not prevent this technology from developing;

– new technologies like blockchain, based on security, authenticity and history, require the preservation and authenticity of certain personal data. Nothing will be able to prevent their progress.

Being in compliance with the GDPR is a necessity in order to continue to be a company with relationships with natural persons (employees, customers, etc.). However, this compliance should not prevent the use of new technologies. On the contrary, these apparent contradictions can give rise to differentiators that can contribute to the sustainability and prosperity of the company.

GDPR innovation must address this issue.

An example of CSR and legal innovation: EcoTree.

It is undeniable that protection of the environment is an increasingly important issue in the eyes of younger generations. For example, the fact that the Europe Ecology party, which came third in the European elections of

May 2019, was well ahead in the 18- to 34-year-old polls, or the buzz surrounding Greta Thunberg's speeches in 2019.

But this strong trend in societies is not limited to the younger generations. Indeed, it now permeates the very traditional world of finance as well. For example, this point is preeminent in the latest letter sent by Larry Finck, BlackRock's CEO, to his clients in January 2020. At the time of writing, BlackRock is the world's largest asset manager, with 6.9 trillion in assets under management. In it, Finck states: "Climate change has become a defining factor in companies' long-term prospects [...] [and] sustainable investing is the strongest foundation for client portfolios going forward."

A good example of innovation in environmental finance entrepreneurship is the company EcoTree.

Conceptually, the principle of this company created by "five childhood friends passionate about the forest" is as follows:

– you buy trees in the forest and become the owner;

– EcoTree either acquires the land itself or enters into a forest lease agreement on land it does not own. Within this lease model, the forest owner receives "in advance" a portion of the revenues expected from logging his or her forest. In return, he or she transfers ownership of a portion of the trees to EcoTree's clients who purchase them. EcoTree takes care of managing these trees, which sequester CO_2 during their decades of growth (this is a long-term "investment" of up to 70 years);

– you can follow the impact of your trees thanks to the different tools and GPS data provided by EcoTree;

– when mature, these trees are cut down and the proceeds go to the original investor, who becomes the owner of the tree (with an expected gain of about 2% per year);

– for each felled tree, EcoTree replants 3.

The strong CSR dimension of this company is further reinforced by the fact that EcoTree has voluntarily set up a very high-level ethics committee.

This illustration is interesting for a second reason, as it is based mainly on a legal innovation. Indeed, the launching of this company was only possible thanks to thorough legal work carried out under the "leasehold estate", which dates back to 1855 and allows for separating the property below (subsoil) from the property above (the trees).

Moreover, it is this legal work that has enabled the company to create one of its main barriers to entry: its registration with the AMF (French Financial Markets Authority). Indeed, at the time of writing, only three French start-ups have been registered with the AMF.

At the beginning of 2020, about five years after its creation, EcoTree manages about 600 hectares of forest and employs more than 30 people. In addition, it has more than 500 partner companies, more than 25,000 individual clients and 40,000 subscribers to its newsletter. In 2019, the company achieved an annual turnover of more than 2 million euros and raised more than 4 million euros in two funding rounds.

4

Financial Evaluation of an Innovation Project

4.1. Introduction

Financial evaluation, or valuation, is a delicate and strategic operation that includes:

– costing;

– goodwill;

– a negotiation.

The financial evaluation of an innovation project (start-up) is essential for third parties to take the decision to invest funds in it.

The start-up is like any other asset: it can be acquired through a purchase transaction.

This acquisition may be total (100% of the shares) or partial (a certain percentage of the total shares) in both cases. To enable a transaction to take place, it must first be valued.

Partial acquisition is the operation most often used by investors to acquire shares of the start-up, not from shareholders, but from the company, in return for a certain amount of funds dedicated to the achievement of the business plan.

The valuation of a company, whether it has potential, like a start-up, or maturity, like an SME, is a delicate operation, because two important milestones must be met:

– a sufficient amount of funds is needed to accomplish the actions required by the business plan;

– the percentage of shares granted in return for the investment must not hinder the development of the start-up by an excess of power, or a demotivation of "key" people, in particular the founders. In general, it is accepted that 60% is an upper limit that should not be reached.

In this issue, the value of the start-up has a central role.

Indeed, for the same investment need, a high valuation entitles the investor to a small percentage of the company's shares, while a low valuation produces the opposite.

From this point on, the investor's objective, always for the same investment, is to have the lowest possible valuation to allow him/her to have the maximum number of shares, while remaining within the limits of the two milestones stated above.

The method for valuation consists of three parts:

– a *costing* based on accounting and financial figures provided by the business plan;

– a *judgment* based on relevant documents that are organized and presented as favorable intangible assets;

– a *negotiation* where the judgment (seen by the investor and to the great displeasure of the other shareholders) is supposed to decrease the theoretical value.

4.2. Costing

Costing is a moment that is especially awaited – and sometimes feared – by the founder(s). Indeed, its very strong psychological aspect cannot be denied, since they know what it will have cost them to get to this point. They have a fair tendency to integrate all the sacrifices made – financial or otherwise – into the initial investment. In their costing, investors should not forget this aspect, otherwise they will come up against a blockage in principle because the founder(s) might not see recognition of their efforts and merits.

Indeed, valuation is often a subject of disagreement between the two stakeholders. Lack of information and lack of trust, as well as asymmetry of information and trust between the parties, generally lead the investor to apply a higher risk premium and claim a larger share of the equity of the firm than the entrepreneur considers reasonable (Douglas *et al.* 2015).

However, from the investor's point of view, it is essential in this first step to rely on the financial component of the latest business plan.

In principle, the "financial" section of the report contains all the figures required for a financial evaluation of the start-up. In particular, it contains a series of operating accounts, accompanied by the investments made and to be made, as well as the cash position. All this information is given over the same period covering the history and the next 3–5 years.

Depending on the information in the business plan, a theoretical value of the start-up can be calculated. For this purpose, numerous well-known methods can be used, for example:

– the net asset method;

– the multiples method (e.g. n times the turnover or m times the margins);

– valuations based on discounted cash flow (DCF method);

– benchmarking valuations.

For a start-up, methods 3 and 4 are the most commonly used.

However, regardless of the method, past or present figures are only a basis for reasoning in support of the credibility of the promises represented by what are commonly referred to as fiscal forecasts (or projections).

The business plan contains the data necessary for calculations. There are three types of data:

– net initial costs;

– current and projected operating and cash flow statements (financial data from the BP);

– business data and different strategies.

The balance sheets, operating and cash flow statements in the business plan must take into account the initial net cost of the project as an expense integrated in the initial situation.

Costing is done in two phases:

– determining net initial costs;

– calculating the theoretical value.

4.2.1. *Determining net initial costs*

The net initial cost is given by the equation (see sections 4.2.1.1 and 4.2.1.2):

Net initial cost = Gross initial cost – Corresponding revenues

4.2.1.1. *Gross initial cost*

It is the sum of the costs (in this case, they are mostly expenses) cumulated from the initial idea to the production of the first heads of series.

When the prototype is ready, the product marketing department must identify one or two customers willing to host a test (beta test). Identifying the host(s) can take a long time.

It should be noted that the cost of a positive beta test is often overlooked. However, it can be of great importance; depending on the information from the customer(s) with whom the test is conducted, a series of more or less important modifications/adjustments can occur (sometimes even calling into question the viability of the project).

Similarly, when a first major order must be delivered, other obligations generate new costs (quality control, documentation, responsiveness, etc.).

The gross initial cost is due to:

– the production of representative samples;

– commercial promotion (promotion, marketing, advertising, etc.);

– the production of the heads of series in accordance with the first orders.

These costs have been used to move the company forward and are therefore considered investments.

4.2.1.1.1. Producing representative samples

About the products:

– material costs are fully expensed, residual inventories are written off;

– the costs of specific non-reusable equipment are fully included in expenses. General-purpose or reusable equipment can impact the cost through depreciation;

– the salary costs charged (all, including management) are fully accounted for;

– overhead costs (space, electricity, etc.) are fully charged.

Concerning the services:

– the costs of specific equipment are fully accounted for;

– general-purpose or reusable equipment can impact cost through depreciation;

– the salary costs charged (all, including management) are fully accounted for;

– overhead costs (space, electricity, etc.) are fully charged.

4.2.1.1.2. Commercial promotion (promotion, marketing, advertising, etc.) and launch

Within this framework, the list of costs incurred is as follows:

– These are mainly costs related to travel, communication (digital, shows, etc.), travel and transportation, intramural visits, and product marketing presentations.

– After a positive beta test, it is particularly important that some customers place a "normal" order to satisfy their real needs. For this, costs for targeted promotions are to be expected. This structuring step shows that the probability that the product/service has found its market is very high.

– After the first significant orders have been received from a few customers, the scale-up phase must begin. This action generates a launch marketing cost (advertising/print/digital, etc.). These costs must be integrated.

4.2.1.1.3. Producing lead products in accordance with first orders

The first customers will be served with products from a top-of-the-range production line. These products are manufactured under non-optimized cost conditions in favor of speed and availability. These cost overruns, compared to the expected standard costs, have to be integrated into the costs of the innovation concerned.

4.2.1.2. *Corresponding revenues*

The costs described in the previous section can generate various kinds of revenue:

– turnover from sales of products/services;

– financing of studies by third parties;

– subsidies, irrevocable and unconditional, directly related to the product/service.

4.2.2. *Calculating the theoretical value of the start-up*

Once the net initial costs have been determined and after any adjustments to the production costs and other expenses, as well as the revenues projected in the business plan, the theoretical value of the start-up can be calculated. One of the methods mentioned in section 4.2 can be used as appropriate.

However, regardless of the method, past or present figures are only a basis for reasoning in support of the credibility of the promises represented by what are commonly referred to as financial forecasts (or projections).

4.3. Judgment (goodwill)

4.3.1. *Purpose*

The judgment is intended to weight the obtained theoretical figure with one of the methods mentioned above. This weighting is justified by a certain number of possibilities of not fulfilling the promises made.

Therefore, it is natural that during the checks, anything that may, to a certain extent, call into question the validity of the financial forecasts and therefore minimize the amount of the value is sought in the relevant documents provided.

4.3.2. *Contents of the data room information source*

To feed the verifications, investors have at their disposal:

– the financial part of the business plan, where all kinds of intellectual constructs will be sought such as inconsistencies, unrealistic figures and unattainable agendas;

– the data room[1] where investors seek to identify, "building their case", any gaps outside of the financial figures. The inevitable presence of gaps must be anticipated, because presented in isolation, the content of this information, which often seems harmless to the start-up, can be worrisome.

4.3.3. *Data room organization and using criteria favorable to innovation*

In addition to the presence of all the documents (commercial contracts, suppliers, customer files, employment contracts with key personnel, leases, etc.) that can prove that the information given in the business plan is complete and sincere, a modern data room must be built to support, or even improve, the theoretical assessment. To this end, the intangible assets of the start-up (HR, profiles, organization, procedures, patents, etc.) must be organized in accordance with criteria favorable to innovation (see section 2.4.2.5), a fundamental mission of a start-up, which reinforces its ability to keep its promises.

A start-up is an organization designed to innovate. Guaranteeing forecasts is an impossibility. On the contrary, creating the right conditions for innovation guarantees the ability to carry out the actions necessary to achieve the expected results.

The eight criteria set out in Chapter 2 must be used, specifying the levels to take into account both the category of the start-up and the level it reaches for each criterion.

1 The place where all the information about the start-up is recorded.

4.3.4. *Using criteria conducive to innovation: approach to be followed*

Using rating according to criteria favorable to innovation (see section 2.4.2.5) allows for developing an argument, not to overvalue the start-up, but to give credibility to the quantified valorization (see section 4.2).

This implies not only the organization of the data room based on the provision of the documents required in section 1.2.1, but also their presentation by the manager (or someone speaking on his/her behalf), explaining the value he/she places on these intangible assets to give credibility to the start-up's ability to achieve its objectives thanks to the skills acquired.

The procedure is as follows:

– categorize the start-up as described in section 1.2;

– in the category, identify the level for each criterion;

– in a criteria/levels table, group the results into scores (e.g. 2/5) and identify strengths and, where appropriate, weaknesses;

– carefully prepare the arguments.

4.4. Negotiation

At a meeting where all the protagonists are present, it is rare that the theoretical costing is questioned. On the contrary, judgments (goodwill or badwill) applied to the theoretical value cause great disappointment among start-ups, because, in general, only the badwill seems to count.

However, this situation is not surprising because, from the investor's point of view, the terms of the initial transaction seem, *a priori*, unbalanced: "We invest funds against promises."

However, the hypothetical can only meet the real through guarantee and trust.

This meeting is first of all the duty of the founder(s) of the start-up. It is their responsibility to create these conditions of guarantee and trust as they progress in their project.

This task must be anticipated as an integral and permanent part of their action. To this end, they must ensure, in parallel with the project relating to their vision, that they carry out a second project, that of its valorization, with the objective of demonstrating that the innovation capacities (transformation of an idea into a product that sells) are present at the time of the transaction.

5

Funding Innovation

A study of 1,053 firms to determine whether there are factors that can influence both the survival and strong growth of start-ups concluded, after extensive analysis, that only three factors really promote the chances of survival and strong growth of start-ups (Cooper *et al.* 1994):

– education level;

– knowledge of the sector of activity in which the entrepreneur embarks;

– initial funding levels.

These results were more recently corroborated by another study of more than 7,000 firms (Dahlqvist *et al.* 2000).

The first two points are self-explanatory and do not necessarily require "technical knowledge" for the founder to act on. On the other hand, initial funding solutions for start-ups have recently become more complex, hence the decision to devote a chapter to this key success factor.

Indeed, Block *et al.* (2017) mention many new types of funding for entrepreneurs that have become (more) available in recent years.

The purpose of this chapter is therefore to recall the issues related to funding innovation before presenting the main funding solutions available to the start-up.

5.1. Issues related to funding growth

For several years now, not a week has gone by without major economic newspapers reporting on a start-up's record fundraising or on the historic amounts raised by venture capital funds.

Fund(ing) and innovation have therefore been intimately linked for several years in the minds of the general public. This link has become so strong that we can sometimes hear, or read, real nonsense that consists of assimilating external funding to revenue.

It seems fundamental to us to recall here, as we indicated in Chapter 1, that a company must above all live off the revenues generated by the sale of its products/services to its customers.

This is obvious and common sense for the majority of experienced entrepreneurs. However, it seems important to remind some entrepreneurs, especially younger ones. The latter may sometimes spend most of their time applying to competitions, or looking for subsidies, at the expense of developing their products or marketing them.

It is crucial to remember that a commercially successful company, with correct margin levels, will eventually have an "internal" cash flow capacity. This is undoubtedly the best way to finance its growth, since it allows the management to maintain control. Moreover, it is significantly less costly than opting for external funding. Finally, a company that is operationally and financially sound will be in a stronger position during negotiations when it needs to call on financial backers.

With this crucial point in mind, it is certain that very fast-growing technology companies will necessarily need external funding. Indeed, in a phase of rapid growth, self-financing will rarely be sufficient to finance the company's numerous projects.

This is all the more true as many start-ups are not profitable during their first years of existence. At the time of launch, the embryonic sales of their first products are rarely sufficient to cover the many expenses and investments necessary at this step.

Beyond the "accounting" profitability, every entrepreneur who has launched a business from a blank sheet of paper will have experienced sleepless nights, and sometimes family tensions, in his or her first years of existence, linked to cash flow problems.

The subject of funding growth is therefore one of the key considerations for the creators of young innovative companies.

Indeed, in the early years of a start-up, the investments required to develop the first prototypes and find the first customers are often substantial. However, even when the company reaches a certain stability, the CEO (or CFO) will have to devote a significant part of their time to find solutions to finance, for example:

– a new production tool;

– growth in working capital requirements. Indeed, in the majority of cases, the company must pay its suppliers and finance its inventory before being paid by its customers. This generates increasing cash requirements as the company grows (which are all the more considerable as the growth is strong);

– significant recruitment of key resources;

– a new version of a product;

– development of new products (and related investments);

– significant marketing campaigns aimed at taking market share from competitors;

– establishment on new geographical markets;

– purchase of competing companies, etc.

In organizational terms, it is important during these growth phases that the start-up retain its entrepreneurial spirit and the degree of innovation of a young company (Hurst 1995).

All studies confirm that funding innovation is a subject in its own right, especially for a fast-growing company.

Indeed, among young innovative companies[1] that file for bankruptcy, more than half of them reach this situation due to lack of funding (Kouame 2012).

However, this question of funding growth is not limited exclusively to the period of company creation.

In fact, in 2015, Raise and Bain & Company identified more than 500 young growth companies that they defined as "companies that have successfully passed through the creation and start-up phases, marketing a product that has reached a defined market, with a 2- to 5-year history and with a strong potential for growth, wealth creation, and employment". They were particularly interested in the difficulties encountered by these entrepreneurs during the company's development phase, i.e. when the company has found, often after a few iterations, a product market pair that seems promising. They define this phase as the "pivotal years". They often occur a few years after the company's registration, and it is during these years that the probability of bankruptcy is highest. For the surveyed entrepreneurs, the first operational difficulty during this phase is recruitment. Financing growth is their second major problem.

In addition, Paulino and Tahri (2014) analyzed the responses given by 175,533 French firms[2] in the (European) Community Innovation Survey on the main obstacles to innovation. Taking all firms together, the two main obstacles mentioned are related to the "lack of financial resources" in the firm and the "cost of innovation" as shown in Table 5.1. These necessary, but not sufficient, conditions for innovation are the weightiest criteria (although they are in the minority with respect to the sum of all others).

This is all the more true when the results are analyzed by focusing only on the innovative companies in this sample, since the percentages of citations of "lack of financial resources" and "excessive cost of innovation" are then four points higher than for all companies, i.e. almost 20%.

1 Young innovative companies or university start-ups are: "New companies that invest in research and development (R&D). They can benefit from tax and social security exemptions." For more information on the eligibility requirements, tax breaks and social exemptions for these companies in France, see https://www.service-public.fr/professionnels-entreprises/vosdroits/F31188.

2 Innovative and non-innovative companies in the manufacturing, service and aerospace sectors.

Indeed, an established company may try to respect the main principles of sound management. It will tend to finance its long-term investments with equity, or long-term debt, and to finance its operating cycle with short-term debt. It may also try to minimize its weighted average cost of capital by seeking to optimize the ratio between its level of equity and debt. However, for the innovative company, especially in its early years, the reality facing the entrepreneur often leaves him with no alternative in terms of its balance sheet structure.

Obstacle group	Obstacles	All companies		
		% of citation	Volume of quotes	Rank obstacle
Knowledge barriers	Lack of qualified personnel	10.3	21,712	4
	Lack of information on technology	3.5	7,366	11
Market barriers	Lack of information on the markets	3.8	8,127	10
	Uncertainty of demand for innovative goods or services	9.8	20,678	5
	No need to innovate because of previous innovations	6.1	12,897	7
	No need to innovate in the absence of a demand for innovation	14.5	30,582	3
Financial barriers	Excessive cost of innovation	15.5	32,744	2
	Lack of means within your company or group	15.7	33,212	1
	Lack of financial means outside of your company	5.9	12,364	8
External barriers	Market dominated by established companies	9.5	20,164	6
	Difficulty finding partners to cooperate with for innovation	5.4	11,393	9

Table 5.1. *Major barriers to innovation*

Indeed, these companies often have limited cash flow and guarantees which, *de facto*, force them to approach certain players depending on their status of progress. We have therefore chosen to describe the means of funding available according to the company's life cycle compared to other more financial approaches (e.g. equity financing versus debt financing).

5.2. Means of funding growth

The following sections summarize the different tools and players likely to contribute to the financing of an innovation according to the stage of the life cycle of the start-up. This "path" is described in a linear and almost exhaustive way, but the reality is often different.

Indeed, it is quite frequent that founders do not succeed in obtaining certain funding and therefore not all of them go through each of these stages. It also often happens that a few funders abound at the same time, because the financing of one of them is conditioned on the contribution of another. Some of the steps described below may therefore be concurrent rather than sequential.

However, the objective remains to present the main methods of financing growth.

5.2.1. *Initial shareholders' equity*

Having a sound project (materialized by the creation of a company) and the significant finances associated with it from the outset significantly increases the company's five-year survival rate as illustrated, for example, in Figure 5.1, from a study carried out by INSEE in the Hauts-de-France region.

Indeed, the higher the initial level of equity capital, the greater the company's chances of long-term survival.

Some founders register their companies with a few thousand euros and sometimes even a few hundred euros.

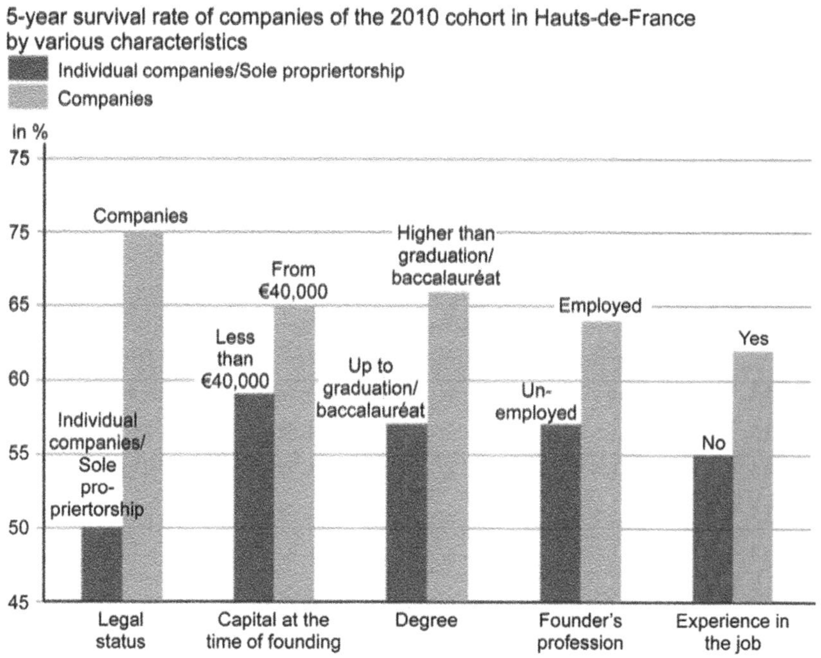

Figure 5.1. *Variation in five-year survival rate by five characteristics[3]. For a color version of this figure, see www.iste.co.uk/dispas/innovation.zip*

If this allows them to create a legal structure, certain points are fundamental and must be kept in mind when building up the initial capital:

– the company's share capital is public in many countries. A company with a share capital of a few hundred euros will therefore be much less credible to third parties (including its potential customers) than a company with a capital of several tens of thousands of euros;

– the level of equity is the cornerstone that will allow leverage (or not) with many of the financing schemes listed in sections 5.2.4, 5.2.6 and 5.2.9;

– the reality is that launching a project (even on a small scale at first) often costs a total of a few tens of thousands of euros. A precise estimate of the investments required for the launch will allow the founder to specify this figure. In the majority of cases, it seems preferable to allocate this sum

3 Source: INSEE (2017). Six entreprises sur dix encore actives cinq ans après leur création. Report.

mainly to the start-up capital rather than registering a company with extremely limited share capital and putting the rest in partner current accounts.

Finally, as stated in the report on the financing of growing SMEs and ETIs carried out in 2015 jointly by several ministries: "Analysis of INSEE's SINE system, that monitors a cohort of newly created companies over five years, shows that among companies that are sustainable for three years, those that have innovated during their first two years of existence will subsequently have an increased probability of strongly increasing their activity. According to this survey, more innovative creators between 2007 and 2009 are notably more likely to have invested during this period. A business creator will therefore have all the more chance of developing his or her business if he or she has quickly initiated an innovative approach by associating it with high investments."

5.2.2. *Love money*

In the majority of cases, the creator also calls on family and friends to help him or her launch the business.

Indeed, resources (henceforth "aid") for the new business are generally mobilized from individuals on the basis of the entrepreneur's previous social relationships (family, friends or professional relationships). The contributions of the donations may require an exchange of financial interest (capital) in the enterprise (Hellmann 2007).

Family, friends or contacts of entrepreneurs can help finance the launch of the company in many ways:

– time and psychological support: this non-financial aspect is fundamental at the time of the launch of a business creation project. In general, the founder will be absorbed day and night by his or her project and will face many pitfalls and doubts. Having family support in this phase is a *sine qua non* for the success of any entrepreneurial project;

– acquisition of a stake in the company: family or friends contribute to the initial capital;

– donations or personal loans: in some countries (and under certain conditions), tax authorities encourage individuals to invest in the creation of a business;

– purchase of the first products or services, for example, during pre-order campaigns organized on crowdfunding sites (see section 5.2.5).

5.2.3. *Competitions*

Business creation competitions are increasingly numerous every year. Some offer no financial endowment, but others can lead to extremely significant funding. In France, for example, we can cite the Ilab competition (organized by the Ministry of Higher Education, Research, and Innovation in partnership with Bpifrance), which provided up to €600,000 in 2019, or the I-Nov competition, which can fund R&D for amounts of up to several million euros. Of course, this type of competition with a significant financial endowment is extremely selective.

The advantage of taking part in these contests is that they oblige founders to formalize the different elements of their project. They also allow them to obtain relevant feedback from experienced entrepreneurs. In addition, they frequently give founders a certain visibility within their ecosystem.

However, it is important to warn younger creators about the time and energy consumed by these contests for returns that are often non-existent. Indeed, many start-upers sometimes spend more time during their first exercise filling out different competition files rather than developing their prototype or interacting with their first users. This is a major risk, since the initial funds are often consumed much more quickly than initially planned. Initial resources, which are often limited, should be largely devoted to fine-tuning the product–market fit.

It is therefore advisable to approach the competitions in a very rational way and with a “return on investment” mindset. Indeed, it is necessary to prioritize and focus only on the few contests that are likely to bring real benefits to the start-up and for which it has objectively reasonable chances of winning a prize.

5.2.4. *Public aid*

Numerous European, national, regional, or local schemes are in place to promote business creation in most European states.

These schemes are not always limited, strictly speaking, to funding a start-up. They may take other forms such as reduced social security charges, tax benefits and loan guarantees.

In France, we can cite, but not exhaustively:

– Pôle Emploi compensation: the State is undoubtedly a very active business angel through this scheme. French entrepreneurs have an unprecedented opportunity to pay for their main personal expenses through Pôle Emploi during the first years of their company's life. This is all the more important since it is rare that founders can afford to pay themselves, via their company, during their first months of business;

– reductions in fees: for example, low-rent accommodation in incubators (partly financed by public authorities). These structures also sometimes grant repayable advances to young companies;

– tax reductions, such as those provided by the JEI status or the research tax credit (CIR).

Once again, what will have to finance the company in the long-run is the income generated by the sales of services/products to its customers and in no case the State. However, it is important to keep two points in mind:

– the financial performance of companies receiving public aid, such as exemptions from social security contributions or investment subsidies, is much better than others. This is true for turnover, profitability and employment indicators. For example, a study of 387 young innovative start-ups monitored over two years[4] shows that companies that have benefited from exemption from social security contributions have seen their sales increase by about 20% per year, compared with 8% for the rest of the sample. In the same vein, start-ups that benefited from investment grants achieved a 16.4% growth rate in profits over the same period, compared with 4.5% for the others;

4 Saty Kouame (2012).

– Bpifrance has done exceptional work in recent years in supporting the creation and development of innovative companies. This is the case in all phases of the life cycle and through countless schemes (from grants to equity investments). This institution has become a *de facto* key interlocutor for entrepreneurs at all stages of financing. Bpifrance's assistance can really get an innovative project off the ground in France.

5.2.5. *Crowdfunding*

Crowdfunding is a relatively new and growing form of funding. It can take three main forms:

– donations, sometimes in exchange for rewards (which may be the first ones the company sells through this means). Pre-sales campaigns are particularly well suited to young companies launching new products;

– loans (which can be relevant, especially for companies that would not be able to obtain them from banks);

– investments (a stake in the company or royalties to be received).

Figures 5.2 and 5.3 illustrate the distribution of the amounts raised by crowdfunding since 2016, in France, by category as well as the average amounts raised.

These funds were raised via the 175 existing platforms in France (Association Financement Participatif France).

Crowdfunding is particularly suited to the early stages of a company's life:

– on the one hand, the average amounts per project in Figure 5.3 are relatively small compared to alternative funding methods;

– 402 million raised through crowdfunding in France in 2018, 80% of total collection volumes funded the growth of VSE/SME/ETI. This was the case in all sectors of activity as shown in Table 5.2.

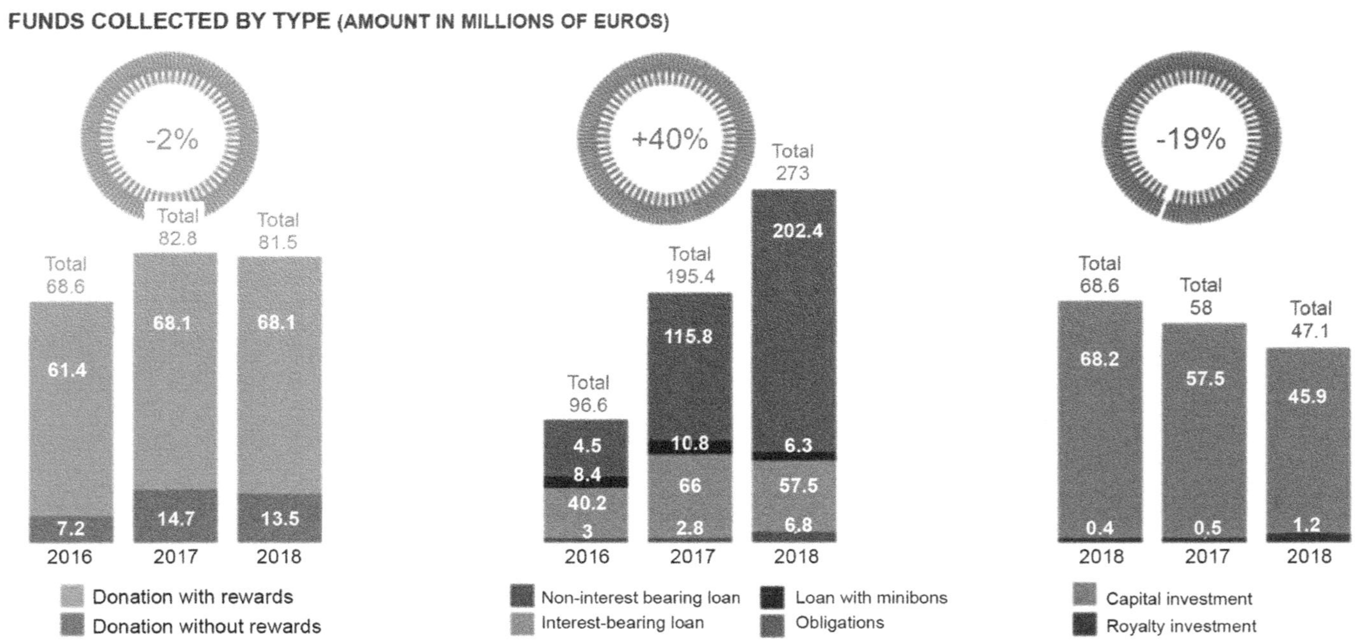

Figure 5.2. *Crowdfunding: funds raised by type (source: "Baromètre du crowdfunding 2018", KPMG and Financement Participatif France, 2018). For a color version of this figure, see www.iste.co.uk/dispas/innovation.zip*

AVERAGE AMOUNT COLLECTED BY PROJECT IN 2018 (IN EUROS)

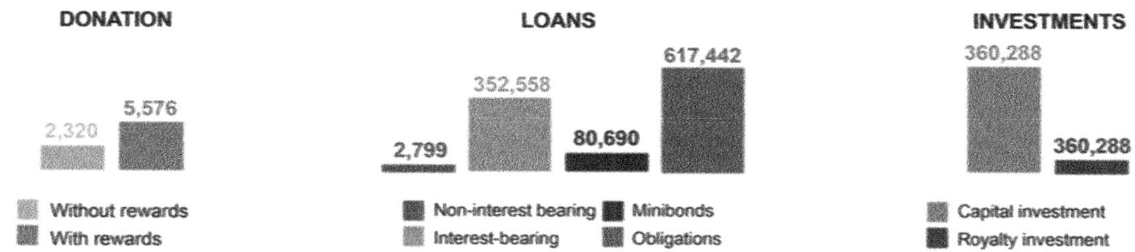

AVERAGE NUMBER OF CONTRIBUTORS PER PROJECT (SINCE THE CREATION OF PLATFORMS)

Donation without reward	Donation with reward	Non-interest bearing loan	Interest-bearing loan	Loan with minibonds	Obligations	Capital investment	Royalty investment
38	76	7	1 041	166	142	83	65

AVERAGE CONTRIBUTION PER PROJECT (IN EUROS SINCE THE CREATION OF PLATFORMS)

Donation without reward	Donation with reward	Non-interest bearing loan	Interest-bearing loan	Loan with minibonds	Obligations	Capital investment	Royalty investment
91€	58€	67€	349€	1 785€	7 504€	10 152€	605€

Figure 5.3. *Crowdfunding: statistics by project (source: "Baromètre du crowdfunding 2018"), KPMG and Financement participatif France, 2018). For a color version of this figure, see www.iste.co.uk/dispas/innovation.zip*

	Donation without reward	Donation with reward	Non-interest-bearing loans	Interest-bearing loans	Obligation	Mini-bonds	Capital investment	Investment in royalties	Total
Plastic arts – live shows	1.2	5.7	0.0	0.0	0.1	0.0	0.1	0.0	7.2
Audiovisual – music	1.0	13.9	0.0	0.7	0.0	0.0	0.0	0.0	15.7
Publishing, journalism, publications	1.1	10.1	0.0	0.5	0.2	0.0	0.3	0.0	12.4
Heritage	2.0	3.7	0.0	0.0	0.0	0.0	0.0	0.0	5.7
Cultural	**5.4**	**33.4**	**0.0**	**1.3**	**0.3**	**0.0**	**0.5**	**0.0**	**41.0**
Agriculture	0.7	5.0	0.9	1.4	1.9	0.1	2.0	0.1	12.1
Shops and services	0.8	9.2	3.4	33.0	2.4	2.7	7.7	0.1	59.4
Environment and ENR	0.8	1.6	0.2	3.3	28.4	2.8	5.6	0.0	42.7
Real estate	0.7	0.9	0.5	2.2	157.7	0.2	13.5	0.0	175.7
Industry	0.7	2.8	1.0	10.4	5.0	0.2	2.3	0.4	22.7
Health and scientific research	0.7	1.0	0.0	3.1	0.6	0.1	3.6	0.0	9.1
Technology and digital	0.7	2.6	0.0	1.9	5.0	0.3	10.1	0.6	21.2
Economic	**5.2**	**23.1**	**6.0**	**55.3**	**201.0**	**6.3**	**44.9**	**1.2**	**342.9**
Teaching, education	0.8	2.7	0.8	0.0	0.0	0.0	0.2	0.0	4.5
Humanitarian and solidarity	1.2	6.3	0.0	0.4	1.1	0.0	0.0	0.0	9.1
Sport	0.8	2.7	0.0	0.4	0.0	0.0	0.4	0.0	4.2
Social	**2.7**	**11.6**	**0.8**	**0.8**	**1.1**	**0.0**	**0.6**	**0.0**	**17.8**
Total	**13.4**	**68.2**	**6.8**	**57.4**	**202.4**	**6.3**	**45.9**	**1.2**	**402**

Table 5.2. *Investments in millions of euros (source: "*La finance participative 4 ans après*", Financement Participatif France, 2019)*

5.2.6. *Interest-free loans*

The interest-free loan is a personal loan granted to the manager without interest or guarantees, which reinforces the company's equity capital. The entrepreneur undertakes to reimburse it under a variable term (generally between two and five years).

There are two main networks in France that grant interest-free loans to entrepreneurs: Initiative France and the Réseau Entreprendre.

The average loan granted to creators in 2018 is about €10,000 for Initiative France and close to €30,000 for the Réseau Entreprendre[5].

Access to bank financing is much easier for business creators when they obtain these interest-free loans. Indeed, joining one of these two networks is a guarantee of quality in itself, which is reinforced by the support provided (by the permanent staff and volunteers of these networks). The consequence is a leverage effect that generally allows the entrepreneur to receive from banks an amount 8 to 10 times higher than the amount of the interest-free loan.

It is also fundamental to note that beyond the funding obtained, the contribution of these networks, particularly through sponsorship, significantly increases the survival rates of companies. They are about 90% at three to five years for these organizations against 60% for the national average according to these networks.

5.2.7. *Business angels*

As the company matures, the entrepreneur will also be able to get closer to business angel (BA) clubs. These investors – former entrepreneurs or executives – have decided to invest part of their savings and experience in start-ups.

By taking a stake in the capital (equity financing), it should be noted that they are in fact involved in the governance, and sometimes in the management, of the companies in which they choose to invest.

---

5 See http://www.reseau-entreprendre.org and www.initiative-france.fr for more details on these networks.

Business angel networks can be grouped by geographical area, by sector of activity or by school. To date, France has 12,000 business angels grouped into 64 networks. These networks have invested approximately €40 million in start-ups in 2018 (France Angels).

BA tickets represent the range above the funding methods presented above, but are lower than the amounts invested by venture capital. Figure 5.4 shows the amounts and the number of operations carried out by the BA in France in 2018. It can be seen that about two-thirds of the tickets were less than €100,000.

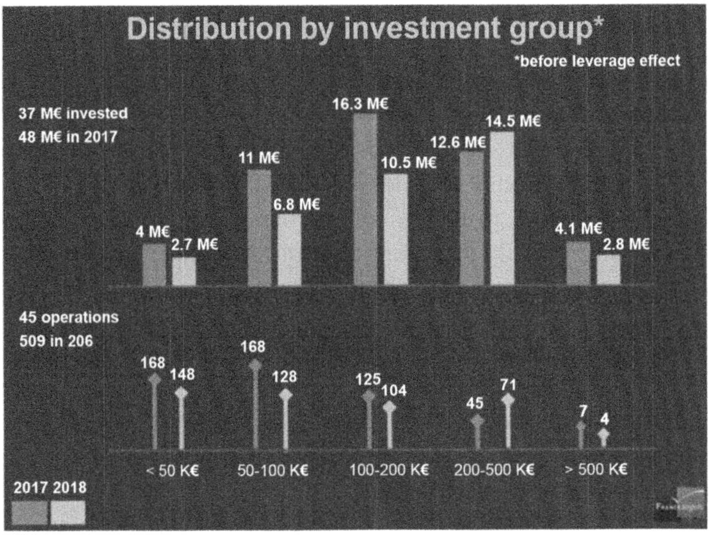

Figure 5.4. *Investments by BAs in 2018*[6]. *For a color version of this figure, see www.iste.co.uk/dispas/innovation.zip*

The contribution of business angels to business performance was analyzed (Kerr 2014). A study has compared, *a posteriori* over a period of six years, the performances of companies that have been funded, in particular by Tech Coast Angels. (the largest BA network in the United States) versus those who have unsuccessfully solicited these business angels.

According to this study, BA-funded companies have 20–25% more chances of surviving the first four years, an additional 20 or so employees, a

6 Source: France Angels (2018). L'investissement des Business Angels en 2018. Document.

17% additional chance of obtaining a patent and a 70% more chances of obtaining other funding.

Admittedly, the companies that were funded may have had more potential for success from the outset (and that is why they were selected).

However, in the same way as for the interest-free loans, the difference in results between the companies financed by these BAs and those that are not is very significant.

From an operational point of view, the entrepreneur can therefore also consider obtaining (or not) BA funding as a signal concerning the potential of his/her project.

It is then a question of finding the right level of intellectual flexibility between not giving up at the first failure, on the one hand, and, on the other hand, not completely ignoring the weak signals that indicate that the project may not have as much potential as the entrepreneur thinks.

In doing so, however, we must take a step back:

– many entrepreneurs, including unicorns, remember that they had to deal with many failures with potential investors before finding new shareholders;

– in Europe, and particularly in France, companies that have been funded by BAs have similar performance to non-BA-funded companies of the same size, age and in the same industry. This is what was revealed from a study conducted on 432 companies funded by French BAs (Levratto, Luc Tessier and Cécile Fonrouge);

– these "angels" can sometimes turn into demons by wishing to interfere in the day-to-day management of the company.

5.2.8. *Venture capital*

As the company continues to grow and develop rapidly, its funding needs to increase. However, its weak assets and still modest financial results often limit the use of traditional banking institutions.

At this stage, venture capital financing, if the start-up can access it at all, is the most suitable method.

It is not always easy to obtain venture capital financing, especially at the seed stage, and this process requires precious time in a world where competition is fierce and accelerating.

Indeed, according to Savignac (2007): "The optimal method of external financing for the firm is an equity contribution from the venture capital firm if the project presents a high risk, if the amount required for its implementation is significant, or if it has a high expected profitability. Conversely, a bank loan is optimal if the risk of failure is low, if the amount of financing is relatively small, or if the firm can provide solid guarantees."

The principle of venture capital is simple: institutional investors such as banks, insurance companies or pension funds entrust part of their funds under management to managers of venture capital companies. These investors take high risks by investing in young innovative companies in the hope of obtaining a significant return on investment from the most successful companies. Statistically, "empirical estimates show that 30% of target companies fail, 40–50% (often called living-deaths) provide a marginal return, while only 10–20% of investments deliver a superior performance" according to Frédérique Savignac (2007).

This model works. For example, at the time of writing, of the top 10 market capitalizations in the world, seven are technology companies. These companies (Alibaba, Apple, Amazon, Tencent, Facebook, Microsoft, Google), which on average are not even 25 years old, were all initially financed by venture capital. Between the years 1975–1977 (when the two oldest technology companies in the top 10 were founded) and the 2000s, the volume of funds granted in the United States by banks to small companies remained stable. At the same time, the funds allocated to venture capital have increased 100-fold (Ueda).

For the entrepreneur, this translates into a much greater capacity for venture capital investment in start-ups than banks. Indeed, the median investment committed by these players per start-up at that time was $4.5 million compared to less than $1 million for banks (Kaplan and Stromberg 2002).

Beyond the amounts invested, venture capitalists also bring non-financial elements to entrepreneurs such as their experience, network or reputation.

Through their presence on the company's board of directors, they also contribute to the company's organizational maturity and "professionalization". Indeed, as Hellmann and Puri point out, venture capital portfolio companies are more effective in hiring sales and marketing VPs, formalizing human resources policies and creating plans to help employees adapt to company growth. It should be noted, however, that these investors are much more likely to replace the company's CEO (even if the founder often remains in the company after this change).

This professionalization also translates into rational operational performance for the company. Indeed, Chemmanur *et al.* (2011) studied 1,881 companies financed by VC in the manufacturing sector between 1972 and 2000. This represented 16,824 years of data and about 20% of all VC-financed companies in the United States. They conclude that the financial performance of VC-financed companies is better than others. This is mainly due to outperformance in terms of revenues. Moreover, these companies are more likely to be listed on the stock market than companies that have not been funded by these players. This conclusion is logical since, by virtue of the mechanism described above, these financiers must "exit" at some point from the capital in order to be able to return their investments (and the majority of the capital gains realized) to their investors.

In simple terms, this exit can be carried out either by reselling the shares to other financiers, by an IPO, or by reselling to an industrial company.

In the first two cases, entrepreneurs are likely to retain their role as CEO of the company. While in the case of a merger–acquisition transaction, the company is often eventually absorbed into a larger competitor and entrepreneurs are unlikely to retain their role of CEO of the company. Indeed, as Schwienbacher (2008) pointed out, this can lead the manager to have an aggressive R&D and innovation policy in order to maximize the chances of listing his or her company on the stock exchange rather than selling it to an industrial player.

In any case, entrepreneurs are more likely to convert all or part of their shares in the company into personal assets if they are funded by venture capital than if they are not. Indeed, whatever the exit method, each offers shareholders a liquidity of their shares.

But personal enrichment is not necessarily the goal of all entrepreneurs. Some will prefer not to open up their capital, grow less quickly and feed their personal assets through regular dividends rather than through the injection of outside capital. This may be an option once the organization has reached a certain stability.

It is difficult to pinpoint a company's size at this point in the funding cycle. This is all the more delicate as venture capital often includes several rounds of investment. However, it is reasonable to estimate that a start-up at this stage of development has passed the size of a VSE (10 people). For example, companies that have benefited from an interest-free loan from Réseau Entreprendre employ an average of 19 people.

However, an econometric study of 18,243 firms (Levratto *et al.* 2011) shows that "a high proportion of equity in the balance sheet is a growth factor for firms with fewer than 20 employees: for small firms, strong equity is essential. The study shows, on the contrary, that it is a factor of decline for larger firms, as if from 20 employees onwards, debt financing becomes more important than equity financing".

It should be noted, however, that a company will, at some point, be limited in its borrowing capacity by the level of its equity capital. Therefore, when it has reached a certain "plateau" in terms of its ability to finance its innovations with debt, it will have to strengthen its equity again.

But at this point, let us focus on bank financing.

5.2.9. *Bank financing*

Start-up entrepreneurs are often critical of banks regarding their low-risk appetite. It is true that a bank must take certain precautions before granting a loan. However, many banks have now launched products adapted to start-ups (sometimes partially guaranteed by the European Union or Bpifrance).

Young entrepreneurs sometimes seem to ignore the banks in their financing methods. However, it is worth briefly recalling that financing growth through debt has at least two significant advantages:

– this financing method is non-dilutive. This characteristic seems to have been forgotten in recent years, but bank financing, when possible, allows the business creator to remain in control. Indeed, recourse to the bank does not lead to dilution of capital (but, in return, the entrepreneur must of course pay back a loan every month);

– interest rates are low. Financing its growth with debt is cheaper than financing it with equity. However, at present, given the historically low levels of interest rates, this is even truer. Indeed, as the Banque de France notes, average interest rates on new loans to non-financial companies have been divided by more than 3 between 2008 and 2018.

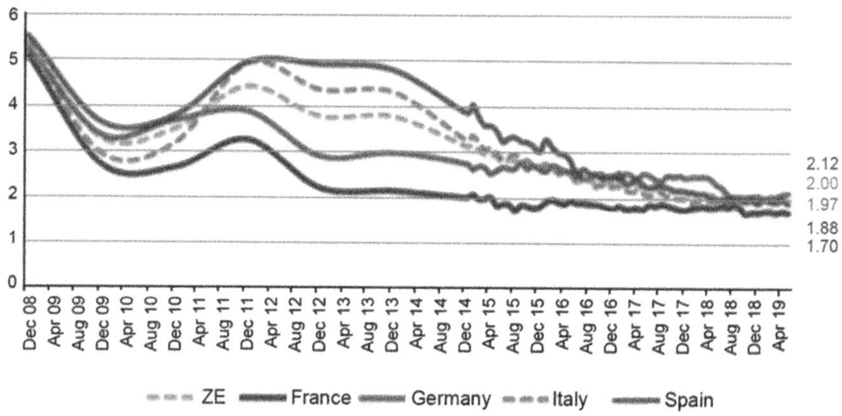

Figure 5.5. *Interest rates for new small business credit (%)*[7]*. For a color version of this figure, see www.iste.co.uk/dispas/innovation.zip*

As already mentioned, while it is not always easy for start-ups to obtain a bank loan, it becomes relatively simple from a certain size upwards.

When the company has succeeded in transforming its invention into innovation, when its products have proven to meet the needs of the market, when customers are willing to pay a price that allows the company to make reasonable margins and when this customer base is growing strongly, the company has turned a corner and naturally turns to banks.

Indeed, for established companies, bank financing is by far the most important source of business financing. This is true for both medium- and long-term financing as well as short-term financing, as shown in Figure 5.6.

7 Source: Banque centrale européenne (2019). Entreprises et PME. Fédération Bancaire Française.

This is the result of a survey, conducted by the European Central Bank, of 11,722 companies, about 95% of which are more than five years old and 90% have fewer than 250 employees.

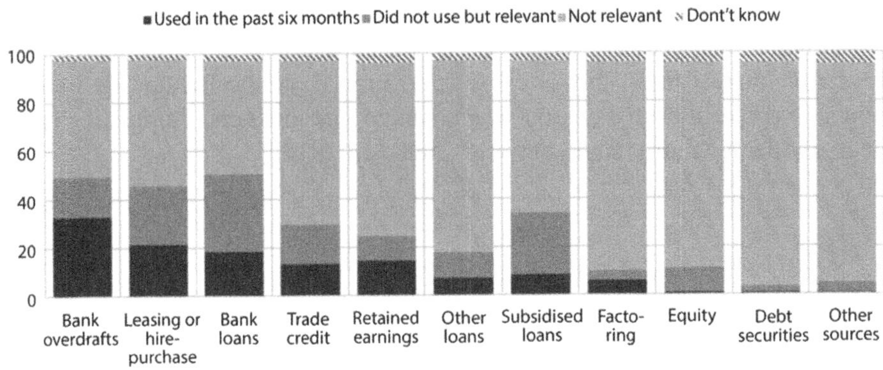

Figure 5.6. *Relevance of different sources of financing for SMEs in the euro area*[8]*. For a color version of this figure, see www.iste.co.uk/dispas/innovation.zip*

According to the same study, while loan take-up rates are a tiny bit lower for European SMEs than for large companies, 73% of SMEs still get the loan they applied for (compared to 83% for large companies).

Established SMEs would therefore be wrong to deprive themselves of this source of financing, which is accessible and a cheaper method of financing than a capital increase.

Going beyond purely financial considerations, it would appear that financing growth through debt reflects sound management decisions: there is a correlation between outperformance and debt levels.

In fact, KMPG selected 1,831 companies (turnover between €10 million and €300 million) that were growing four times faster than the average growth of their industry for five years. The purpose of this study was, among other things, to identify their growth drivers. These outperforming companies, which represent 10% of French companies, use two main levers: internationalization and debt. "Their debt-to-equity (gearing) ratio is high, at 61% in 2006, almost double that of all family-owned companies. It is

8 Source: European Central Bank (October 2018–March 2019). Survey on the Access to Finance in the Euro Area.

therefore close to that of CAC 40 and SBF 250 companies and much higher than that of family-owned SMEs."

Of course, it is not necessarily the level of indebtedness that "drives" the good operating performance of these companies. These companies have good products, a stable customer base and comfortable margin levels generating recurring cash flow. This enables them to look to the future with confidence and to launch new projects. Thanks to their strong balance sheets, part of these projects can be financed by debt.

However, if the company continues its exponential growth for many years, significant exceptional financing needs will arise. This could be, for example, to finance the construction of a new plant, the acquisition of a company, a diversification, or an international expansion.

The fast-growing company will then not be able to finance everything with debt. In most cases, management will have to recapitalize the company in order to maintain a reasonable debt-to-equity ratio (gearing).

A company whose funding would be based exclusively on loans would run the risk of being suffocated by its financial expenses if its operating performance suddenly deteriorated. This can happen, for example, when market needs change, when a new technology or a new competitor arrives.

This is one of the reasons why bankers monitor a company's debt-to-equity ratio (debt/equity). At a certain level of indebtedness, a company that wants to continue to fund its investments and growth will have no choice but to increase its equity.

It is at this point that the company will consider (re)opening up its capital and may call on so-called "equity financing" investors to strengthen its equity capital.

5.2.10. *Private equity*

When a company reaches a certain level of maturity and its cash flows are stabilized, it can call on equity investors to fund its growth. From the point of view of institutional investors, these players are part of the same asset class as venture capitalists. By way of comparison, the funds managed by this asset class represent, at the time of writing, approximately 80% of the total amount of gold reserves in the world. As illustrated in Figure 5.7, private markets manage about $6 trillion against an estimated $7 trillion in gold value.

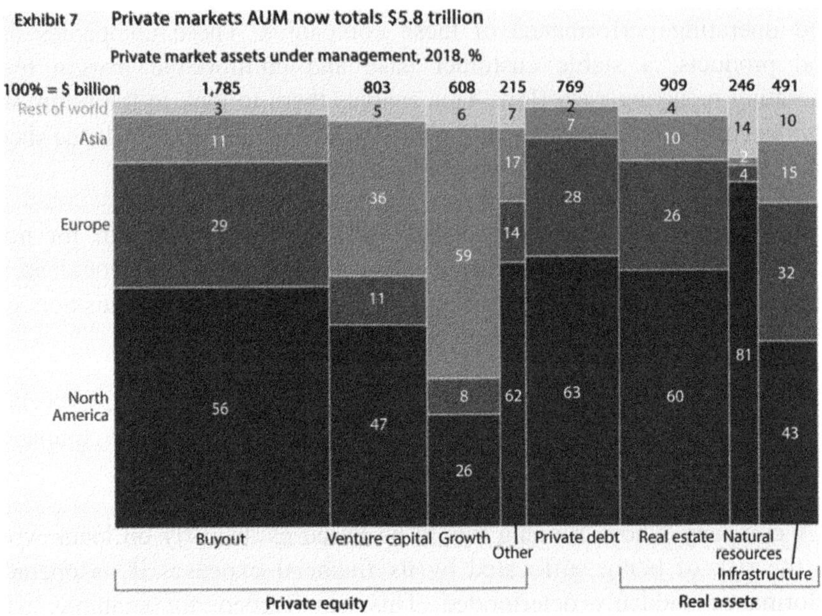

Figure 5.7. *Assets under private market management (Bökberg* et al. *2019). For a color version of this figure, see www.iste.co.uk/dispas/innovation.zip*

The operating principle of private equity players is exactly the same as that of venture capital players. They invest in companies with the expectation of obtaining, after 5 to 10 years, returns on investment for their shareholders that are much higher than those that the latter could have obtained on much less risky asset classes (e.g. bond markets).

It is possible to call on these equity investors as soon as the company has a certain financial stability, since these players very often use the leverage effect of debt in parallel with their entry into the capital.

For the entrepreneur, this option can be a good prerequisite or even an alternative to an IPO.

Indeed, according to the McKinsey report (Bökberg *et al.* 2019), the value of portfolio companies of private equity funds is now higher than the market capitalization of listed companies.

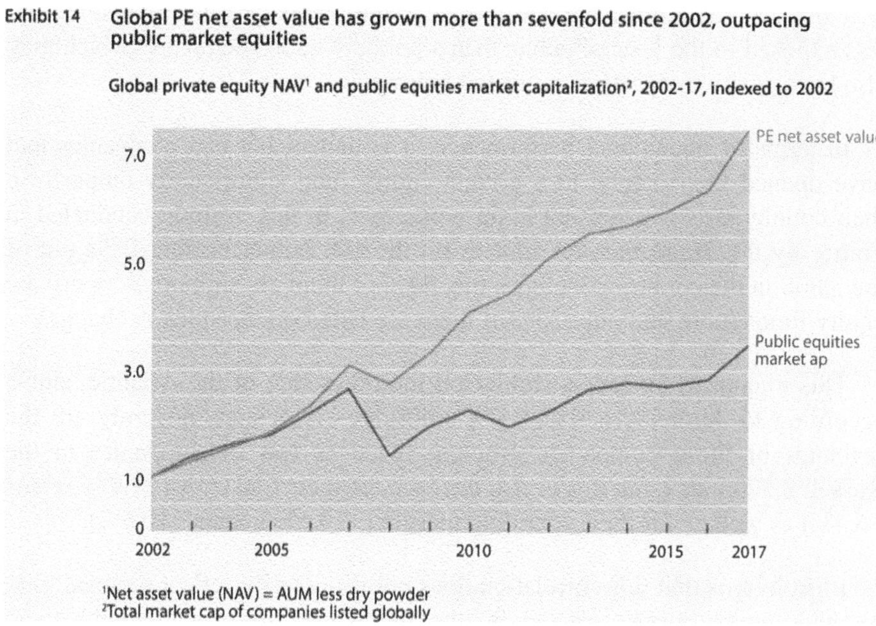

Figure 5.8. *Comparison of equity market valuations and private funds (source: McKinsey Report). For a color version of this figure, see www.iste.co.uk/dispas/innovation.zip*

There are some advantages to using these funds for the entrepreneur:

– funding growth with amounts that can be very significant, in particular by using debt leverage;

– implementing formalized governance and reporting procedures. This can be an advantage or a disadvantage by the entrepreneur. However, it is important to keep in mind that the interests are aligned and that the objective of the equity investor is the success of the company in which he or she invests in. It is also important to note that these procedures are much less burdensome than those imposed by an IPO (and the financial communication that is necessarily linked to it);

– have a shareholder at their side who invests for the long-term: PE investors keep their holdings for more than five years on average (see Bökberg *et al.* (2019));

– these funds are private and therefore the company (and its results) are less exposed to the general public than a publicly traded company (which may also have positive aspects as indicated below).

In terms of operational performance, it is undeniable that companies that have opened their capital to a private equity fund significantly outperform their counterparts who are not in the same case. In fact, a study conducted in France by the *Magazine des Affaires* on the 190 outperforming ETIs out of the 5,800 in the country concludes that 84% of them are supported by private equity funds (note that only 20% of them are listed on the stock exchange).

This phenomenon seems similar on the other side of the Atlantic, since, according to Andrea Auerbach of Cambridge Associates, not only are the revenues of these companies growing twice as fast as companies in the Russell 2,500 index, but this is also the case for their EBITDA (+7.9% versus +6.4%) as well as for their operating margin (+19.5% versus +10.6%).

It is obvious that this correlation does not necessarily reflect a causal link. In other words, firms selected by the funds may intrinsically have more potential, even before equity investment, than their non-funded counterparts. It is interesting to note, however, that the entrepreneur is more likely to generate "abnormal returns" if a portion of the capital is open to investment funds.

This characteristic persists even if the fund exits, in whole or in part, from the capital via an IPO. Indeed, a study was conducted on 1,595 IPOs in the United Kingdom (Levis 2011) to compare the performances of these companies according to their initial shareholding. Of these companies, 1,141 were not supported by a fund, 250 were supported by a venture capital fund and 204 were supported by a private equity fund. This study concludes that three years after the IPO, the stock market performance of the companies that were supported by a private equity fund outperformed the others.

One of the criticisms sometimes leveled at equity investors is that this improvement in financial performance comes at the expense of innovation and that it favors increasing margins in the short-term. However, a U.S. study that analyzed the patent filings of 472 companies with private equity

(Lerner *et al.* 2011) concludes that not only does the pace and quality of innovation not dry up following the fund's investment, but also, on the contrary, the patent portfolio becomes more "focused" following the fund's investment. This also seems to be the case in Europe, since an econometric study by the European Central Bank concludes that European private equity contributes to 8% of industrial investments, but 12% of industrial innovation.

As with venture capital, the entrepreneur must be aware that private equity investors will seek to increase the value of their participation in the long-term and thus to exit the capital of the company they have entered.

In the majority of cases, this resale is made to a strategic player, as illustrated in Table 5.3, which summarizes the exits made in the context of more than 17,000 LBO transactions.

Year of original LBO	*1970–1984*	*1985–1989*	*1990–1994*	*1995–1999*	*2000–2002*	*2003–2005*	*2006–2007*	*Whole period*
Type of exit:								
Bankruptcy	7%	6%	5%	8%	6%	3%	3%	6%
IPO	28%	25%	23%	11%	9%	11%	1%	14%
Sold to strategic buyer	31%	35%	38%	40%	37%	40%	35%	38%
Secondary buyout	5%	13%	17%	23%	31%	31%	17%	24%
Sold to LBO-backed firm	2%	3%	3%	5%	6%	7%	19%	5%
Sold to management	1%	1%	1%	2%	2%	1%	1%	1%
Other/unknown	26%	18%	12%	11%	10%	7%	24%	11%
No exit by Nov. 2007	3%	5%	9%	27%	43%	74%	98%	54%
% of deals exited within								
24 months (2 years)	14%	12%	14%	13%	9%	13%		12%
60 months (5 years)	47%	40%	53%	41%	40%			42%
72 months (6 years)	53%	48%	63%	49%	49%			51%
84 months (7 years)	61%	58%	70%	56%	55%			58%
120 months (10 years)	70%	75%	82%	73%				76%

Note: The table reports exit information for 17,171 worldwide leveraged buyout transactions that include every transaction with a financial sponsor in the CapitalIQ database announced between 1/1/1970 and 6/30/2007. The numbers are expressed as a percentage of transactions, on an equally-weighted basis. Exit status is determined using various databases, including CapitalIQ, SDC, Worldscope, Amadeus, Cao, and Lerner (2007), as well as company and LBO firm web sites. See Strömberg (2008) for a more detailed description of the methodology.

Table 5.3. *LBO exit solutions by period of initial investment (source: Kaplan and Stromberg (2009))*

As mentioned above, this often results in the entrepreneur losing management of his or her company in the more or less short-term. Indeed, the latter is then absorbed into an industrial player often of a much larger size than the target. The exit for the fund may consist of selling to another fund (secondary buyout in Table 5.3) and the same question will then arise for the entrepreneur in the long-term. The only viable solution to stay in control is to push for an exit via an IPO (as explained in section 5.2.8 on venture capital).

In addition to these benefits for the individual entrepreneur, the stock market has many other positive factors specific to the entrepreneur:

– it offers real liquidity to the portion of the company's capital that is listed on the stock market (free float);

– it allows a great visibility and offers a certain guarantee of seriousness with regard to third parties (especially the general public);

– it gives a market value to the company.

As previously mentioned, the main disadvantages of an IPO are the high costs associated with it and the short-term pressure generated by shareholders.

6

Valorization of Innovation by Financial Markets

In the first chapters of this book, we discussed the very concept of innovation and what it means for entrepreneurs and businesses. On this basis, we will turn the problem upside down and seek to better understand whether the effective management of innovation by companies is recognized and valued by financial markets. Is it favored or, on the contrary, hindered by constraints such as risk aversion, information asymmetry or focus on short-term results?

In this chapter, we will attempt to better understand the relationship between innovation and financial market performance. Initially, we will focus on the reaction of the stock markets. Next, we will analyze the relationship from the perspective of the fixed-income markets.

To do so, we will first go back to the fundamental concepts for analyzing the performance of equity markets. We will have to resort to the use of valuation models for financial assets. These models allow us to better understand the evolution of stock market returns.

The analysis of the performance of innovative companies by these valuation models should thus enable us to answer the central question of this section: are innovative companies well recognized and rewarded by the financial markets?

6.1. How to determine the expected return from innovative companies

In order to be able to determine the most accurate possible quantification of the return expected by investors on various financial assets over a given time horizon, it is necessary to use models. Models have been in constant development for more than half a century and continue to evolve in order to get closer and closer to the behavior of stock markets.

6.1.1. *Capital asset pricing model (CAPM)*

This model is undoubtedly the one that has had the greatest impact on portfolio management. Developed in 1964 by William Sharpe, this model provides a practical solution to the question of determining expected returns.

Once again, the objective is to popularize these concepts and not to enter into technical developments. We will exclusively try to illustrate the practical usefulness of such a model for investors.

The basic principle of this model is that the main risk for investors is the so-called "market" risk. The way the market will react to macroeconomic, political, monetary or other developments will drive the vast majority of individual stocks.

It is very important not to confuse expected return with actual return. Here, we will estimate the expected return, i.e. estimate what "should" happen. Since we do not know the future, it is only on this basis that investments can be determined.

In reality, the return received may be quite different from what was initially expected, as many unforeseen elements, surprises, may occur during the investment period. The idea is therefore that "excluding surprises", the model gives a good approximation of what the return should be.

The model is defined as follows:

$$E(R_i) = R_f + (R_m - R_f).\beta$$

The expected return on a share is thus defined as the sum of the risk-free rate (R_f) and the market risk premium ($R_m - R_f$) adjusted for the sensitivity of the share to the market (β). In other words, if the share is more sensitive, or even riskier than the market, it will rise more when the market rises and, of course, will also fall more when it falls. Graphically, it is clear that the model reduces the problem to two dimensions whose relationship is linear: the expected return and the expected risk. The line that connects these different points is called the security market line (SML).

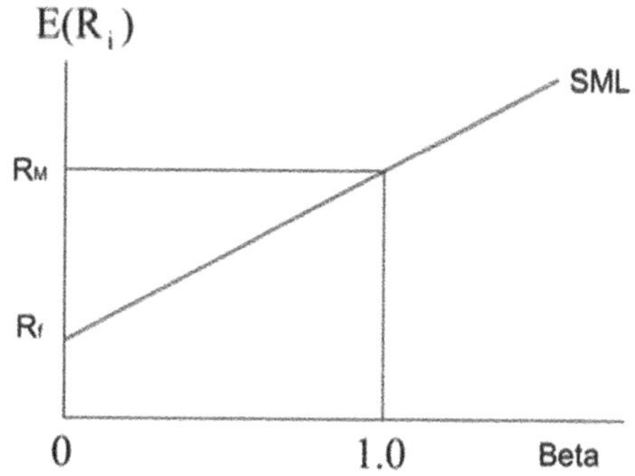

Figure 6.1. *Relationship between a share's expected return (E(Ri)) and its risk (β) (inspired by Reilly and Brown (2012)). For a color version of this figure, see www.iste.co.uk/dispas/innovation.zip*

This model has had, and still has, a central impact in portfolio management. It is therefore quite common to hear portfolio managers say that they favor "high beta" stocks (highly sensitive) or low beta depending on their market anticipation. This has become common parlance. If a stock has a beta of one, it should have a sensitivity, or risk, equal to that of the market.

Within the context of our work, the idea is therefore to see, on the basis of such a model, whether innovative companies are favored by the financial markets.

To do so, we will use a commonly used tool: Microsoft Excel. Once again, the idea is to popularize and make accessible to all the basic tools for the analysis of financial assets.

The innovative companies we will analyze were selected based on a 2019 study by the Boston Consulting Group.

For reasons of consistency of results, we will focus solely on US companies, but our analysis can of course be extended to all markets.

EXHIBIT A | 2019 Most Innovative Companies

1	Alphabet/Google	11	Boeing	21	McDonald's	31	AXA	41	Dell
2	Amazon	12	BASF	22	Marriott	32	Unilever	42	Walmart
3	Apple	13	T-Mobile[2]	23	Alibaba	33	Salesforce	43	eBay
4	Microsoft	14	Johnson & Johnson	24	Bayer	34	Pfizer	44	HP Inc.
5	Samsung[1]	15	DowDuPont	25	AT&T	35	Stryker	45	ING
6	Netflix	16	Siemens	26	Allianz	36	NTT Docomo	46	BP
7	IBM	17	Cisco Systems	27	BMW	37	Toyota	47	Daimler[4]
8	Facebook	18	LG Electronics	28	SAP	38	Volkswagen[3]	48	Huawei
9	Tesla	19	Vale	29	Philips	39	3M	49	Rio Tinto
10	Adidas	20	JPMorgan Chase	30	Royal Dutch Shell	40	General Motors	50	Hilton

[1]Includes only Samsung Electronics
[2]Includes only US T-Mobile, not Deutsche Telekom
[3]Includes Audi and Porsche
[4]Includes Mercedes-Benz

Figure 6.2. *Top 50 of BCG's 2019 most innovative companies of BCG (source: Ringel* et al. *(2019)). For a color version of this figure, see www.iste.co.uk/dispas/innovation.zip*

We will thus analyze the 20 American companies considered as the most innovative by the BCG study. We have chosen the top 20[1] in order to ensure a certain sector diversification. We have included the GICS (Global Industry Classification Standard) sectors in Table 6.1.

We calculated the "beta" of each of these companies in relation to the S&P 500 Index, which is a benchmark for the U.S. market and is probably the most widely used index for analyzing companies with large market capitalizations.

We have used the longest possible study period on these (weekly) data, namely since the IPO of Facebook (May 25, 2012 to November 29, 2019).

1 We have omitted DowDuPont with regard to the capital operations of the last few years.

	Name	GICS sector
1	Alphabet/Google	Telecommunications
2	Amazon	Consumer Discretionary
3	Apple	Information Technology
4	Microsoft	Information Technology
5	Netflix	Telecommunications
6	IBM	Information Technology
7	Facebook	Telecommunications
8	Tesla	Consumer Discretionary
9	Boeing	Industry
10	US T-Mobile	Telecommunications
11	Johnson & Johnson	Health
12	Cisco Systems	Information Technology
13	JP Morgan Chase	Finance
14	McDonald's	Consumer Discretionary
15	Marriott	Consumer Discretionary
16	AT&T	Telecommunications
17	Salesforce	Information Technology
18	Pfizer	Health
19	3M	Industry
20	General Motors	Consumer Discretionary

Table 6.1. *Top 20 most innovative U.S. firms and their industry (source: Bloomberg and BCG (2019))*

EXAMPLE.– Calculate the beta of the CAPM model using Excel.

To calculate the beta of the securities using Microsoft Excel, we used the following formula:

=INDEX(LINEST(excess return[2] of the share to be analyzed;excess return SP500;;TRUE);1;1)

2 Excess return refers to the difference between the stock return and the risk-free rate.

The LINEST function refers to the results of a linear regression in Excel "1;1" refers to the first coefficient starting from the end of the regression. We could also have used the function =SLOPE (excess return of the share to be analyzed; excess return SP500), but this function is less powerful. Indeed, it is suitable for the beta of a one-factor equation, but not in the case of a multifactor equation. Moreover, the LINEST function allows us to not only study the sensitivity coefficients of the different variables, but also the standard errors. This is very useful because it allows us to calculate Student's t-statistic to determine the significance of the result[3].

The standard error is calculated using the same function, but with reference to ";2;1", ";2;2", etc. For example, for the first share of our analysis, Alphabet/Google, the t-statistic is calculated by dividing the beta by the standard error, i.e. in Excel:

> =INDEX(LINEST(excess return Alphabet/Google;excess return SP500;;TRUE);1;1) /INDEX(LINEST (excess return Alphabet/Google;excess return SP500;;TRUE);2;1)

In our sample, the Alphabet/Google beta is 1.16 and the t-statistic is 15.68. For simplicity, a t-statistic result above 1.96 will be considered positive and significant and below -1.96, negative and significant. In this example, the beta CAPM, i.e. market sensitivity, seems to be a very significant variable to explain the return of Alphabet/Google. It is 1.16; so, the stock was more sensitive (or risky) than the market over the period studied.

Finally, it is also possible to calculate the "alpha" of the share, i.e. the return that would not be explained by the risk. This is, of course, what investors are looking for, i.e. securities that "outperform" the market beyond the risk taken. A very innovative company, if this is not sufficiently taken into account by the market, could thus deliver significant positive alpha.

To calculate the alpha using Excel, we need to regress the excess return of the stock (i.e. stock return minus the risk-free rate) on the excess return of the benchmark (i.e. SP500 return minus risk-free rate). The constant of the regression gives us the alpha.

---

3 For more details on t-statistics, please refer to a book on statistics. For the sake of popularization, a result greater than 1.96 indicates that there is a 95% chance that the coefficient under review is positive and statistically different from zero. If the result is less than -1.96, there is a 95% chance that the result will be negative and statistically different from zero.

In Excel, we will once again use the function LINEST, but pointing to the coefficient ";1;2", which is the second coefficient from right to left.

=INDEX(LINEST(excess return[4] of the share to be analyzed;excess return SP500;;TRUE);1;2)

The result is 0.11% on a weekly basis and 5.78% on an annual basis[5].

And to calculate the t-statistic:

=INDEX(LINEST(excess return Alphabet/Google;excess return SP500;;TRUE);1;2) /INDEX(LINEST(excess return Alphabet/Google;excess return SP500;;TRUE);2;2)

In our example, the result is 0.87 for this statistic, not significantly different from zero. It is interesting to note that an alpha of 5.78% per year, a level that appears relatively high, is not statistically significant. This can be explained by the high volatility of the stock.

It appears that the answer to our question about the stock market performance of innovative companies is not so obvious. Indeed, a higher historical performance can be perceived as a company whose evolution has positively surprised the markets by its prospects or as a higher risk. This is where the use of a valuation model is very useful. Indeed, if the performance is essentially explained by the risk, there will be no "alpha", i.e. a positive return not explained by the risk. When we regress the excess return ($R_i - R_f$) of our innovative companies on the excess return of the market ($R_m - R_f$):

$$R_{it} - RFR_t = \alpha + \beta.(R_{mt} - RFR_t) + e_{it}$$

The beta (β) represents the systematic risk of the share. If we are above 1, the stock is riskier than the market, and if we are below 1, it is less risky. The alpha (α) represents the return not explained by the risk. If it is significantly positive, it means that the company has recorded a return in excess of its risk over the period studied.

The results are shown in Table 6.2.

4 Excess return refers to the difference between the shareholder return and the risk-free rate.

5 Since our returns have been calculated on a continuous basis, in order to go from weekly returns to annual returns, you only need to multiply by 52: 0.11% x 52 = 5.78%.

	Beta CAPM	Rank	Annualized standard deviation	Rank	Annualized return	Rank	Annualized alpha	Rank
Netflix	1.58	1	50.30%	1	45.64%	1	27.11%	1
Tesla	1.14	8	46.74%	2	32.85%	2	19.53%	2
US T-Mobile	0.84	17	30.65%	4	24.03%	4	14.18%	3
Amazon	1.26	4	26.96%	6	28.19%	3	13.41%	4
Facebook	0.98	14	35.18%	3	22.00%	6	10.46%	5
Boeing	1.03	13	23.22%	9	22.05%	5	9.98%	6
Microsoft	1.10	9	21.09%	13	21.74%	7	8.85%	7
Alphabet/Google	1.16	7	23.02%	10	19.42%	9	5.78%	8
Marriott	1.06	12	21.17%	12	17.70%	11	5.30%	9
Salesforce	1.33	2	27.40%	5	19.83%	8	4.28%	10
JPMorgan Chase	1.20	6	20.68%	14	18.12%	10	4.09%	11
Apple	1.08	11	25.62%	8	16.68%	12	3.98%	12
McDonald's	0.57	20	15.16%	19	10.22%	15	3.50%	13
Johnson & Johnson	0.74	18	14.52%	20	10.25%	14	1.57%	14
Cisco Systems	1.20	5	22.30%	11	13.39%	13	– 0.65%	15
Pfizer	0.87	16	18.10%	17	7.07%	17	– 3.18%	16
3M	1.09	10	18.41%	16	9.39%	16	– 3.36%	17
AT&T	0.72	19	17.44%	18	1.39%	19	– 7.08%	18
General Motors	1.30	3	26.64%	7	7.02%	18	– 8.26%	19
IBM	0.94	15	20.03%	15	– 4.98%	20	– 15.99% (*)	20
S&P 500	1.00		12.97%		8.45%			

(*) Significant at a 5% threshold.

Table 6.2. *Ranking based on annualized alpha of the 20 most innovative US companies (source: Bloomberg and BCG (2019))*

So, it seems that the level of maturity of the innovative company is very important. Younger companies (Netflix, Amazon, Salesforce, etc.) tend to have higher betas, while established companies (McDonalds, AT&T, etc.), although still very innovative, have lower expected performance. However, some results are surprising, for example, Facebook with a beta of 0.98. On the other hand, if we analyze the risk based on standard deviation, all these companies show a relatively high risk (from 20% for IBM to 50% annual volatility for Netflix). It should be noted that none of these companies show an alpha significantly different from 0 at a 5% threshold, even though the performance was very good. The volatility, or risk, of these securities therefore seems to explain the high returns over the period by many of them.

The CAPM, with its market factor alone, seems limited to answer our question. Summarizing all the risks of innovative companies to market sensitivity does not make it possible for us to draw a clear conclusion. Moreover, apart from our problem, the CAPM shows its limits in explaining the (out)performance of many share classes such as small market capitalizations, value shares (Fama and French 1992) or the momentum effect (Jagedeesh and Titman 1994).

For these different reasons, we will move on to the analysis of multi-factor models that will allow us to make a more refined analysis.

6.1.2. *Multifactorial models: arbitrage pricing theory (APT)*

Faced with the limitations of the CAPM, many multi-factor models have been developed to try to better explain stock returns. From a theoretical point of view, the most successful is the arbitrage pricing theory (APT), developed by Stephen Ross in 1976 and further developed in 1980 by Richard Roll and Stephen Ross. The authors showed that a single-factor model could not explain all returns on financial assets and that it is necessary to move on to several factors. The expected return on a stock is thus equal to the sum of the product of the risk premiums linked to each of these factors and the sensitivity of the stock (or portfolio) studied to each of these factors:

$$E(R_i) = p_0 + b_{i1}.p_1 + \ldots + b_{ij}.p_j$$

Unfortunately, the model does not define the factors that explain the returns of the shares. In an applied version of this theory, Chen *et al.* (1986) developed a model based on various macroeconomic factors such as industrial production and inflation, to which they add the risk premium of risky bonds over the risk-free rate, the slope of the yield curve and the market risk premium.

Like the results obtained for the CAPM, these types of macroeconometric models do not provide us with much more information, with once again two types of innovative companies emerging: mature companies that appear less risky and younger companies that are more at risk. Moreover, these models still have difficulty explaining the differences in returns between value and growth stocks, for example.

The APT thus brings a crucial element by legitimizing the use of multifactorial models, but it is relatively difficult to set up and poorly adapted to our problems.

6.1.3. *Multi-factor models based on microeconomic factors*

In 1992 and 1993, Fama and French (FF) brought a major breakthrough with their three-factor model:

$$R_{it} - RFR_t = a_i + b_{i1}.(R_{mt} - RFR_t) + b_{i2}.SMB_t + b_{i3}.HML_t + e_{it}$$

with:

– SMB = small minus big;

– HML = high (book-to-market) minus low.

By adding the SMB and HML factors, Fama and French developed a model of great simplicity, as well as a formidable efficiency. Thus, the vast majority of equity returns can be explained by their sensitivity to the market risk premium, by the risk premium of small-cap stocks (which reflects the liquidity risk associated with this type of investment) and the value risk premium, i.e. the premium observed over the long-term for low-valued stocks (through a price-to-book ratio[8] in the FF model) compared to so-called "expensive" stocks. FF justified this outperformance by a default risk premium, the cheapest shares having, according to them, a greater risk of default.

EXAMPLE.– Developing the Fama and French model using Excel.

To analyze stock performance using a Fama and French three-factor model, the principle is identical to that developed for the CAPM. It is therefore necessary to regress the excess return of the share to be analyzed on the return of the three factors of Fama and French[9].

To perform this regression, we use the LINEST function again:

=INDEX(LINEST(excess return of the share to be analyzed;return of the 3 FF factors;;TRUE);1;1)

Based on the above equation, ";1;1" will represent the coefficient of HML. A value share will have a positive and statistically significant coefficient. A growth share will have a negative and statistically significant coefficient. If the coefficient is not significant, nothing can be concluded.

Note that it is necessary to select the three factors together as explicative variables. Moreover, since there are three factors, it is necessary to calculate a formula for each explanatory variable (here, the explanatory variables are the Fama and French factors). "1;1" points to the "last" factor, ";1;2" the previous one and so on. So, we go from right to left on the equation, the last factor being ";1;1".

For SMB:

=INDEX(LINEST(excess return of the share to be analyzed;return of the 3 FF factors;;TRUE);1;2)

A small share will have a positive and statistically significant coefficient. A big share (i.e. a large market capitalization) will have a negative and statistically significant coefficient. If the coefficient is not significant, nothing can be concluded.

For RMRF:

=INDEX(LINEST(excess return of the share to be analyzed;return of the 3 FF factors;;TRUE);1;3)

---

8 Price per share divided by book value per share.

9 These are freely available on Kenneth French's website at the following address: http://mba.tuck.dartmouth.edu/pages/faculty/ken.french/datalibrary.html.

For the alpha:

=INDEX(LINEST(excess return of the share to be analyzed;return of the 3 FF factors;;TRUE);1;4)

For the latter, do not forget that if the data is weekly, the alpha will also be weekly. It is therefore advisable to annualize for a clearer reading of the result.

The microeconomic approach produces a type of model that is much closer to the actual behavior of markets.

	Alpha 3 FF factors, annualized	**Style**
Netflix	22.55%	Growth (*)
Tesla	17.20%	Growth (*)
US T-Mobile	12.33%	
Amazon	9.04%	Growth (*)
Boeing	7.40%	
Facebook	6.84%	Growth (*)
Microsoft	4.74%	Growth (*)
JPMorgan Chase	4.02%	Value (*)
Marriott	3.74%	
Alphabet/Google	1.40%	Growth (*)
McDonald's	1.40%	Growth (*)
Apple	0.11%	Growth (*)
Salesforce	– 0.33%	Growth (*)
Johnson & Johnson	– 0.66%	
Cisco Systems	– 3.81%	
3M	– 5.50%	Value (*)
Pfizer	– 6.00%	Growth (*)
AT&T	– 8.44%	Value (*)
General Motors	– 8.85%	Value (*)
IBM	– 17.81% (*)	

(*) Significant at a 5% threshold.

Table 6.3. *Ranking based on annualized alpha of the 20 most innovative US companies and their management style calculated using the FF three-factor model (source: Bloomberg and BCG (2019); Kenneth French website)*

An analysis of innovative companies by these three factors (Table 6.3) shows that many innovative companies have the characteristics of growth companies, particularly the most recent companies active in the new technologies sector (Netflix, Tesla, Amazon, etc.). However, we also observe that other companies, all of which are more mature, such as AT&T or JP Morgan, have value characteristics. Finally, this analysis does not inform us about six companies out of 20, the results not being statistically significant.

On this basis, we might be tempted to conclude that innovative companies tend to be growth companies according to the FF approach. However, as Fama and French illustrated in their 1992 article, growth companies have shown a lower return than value companies over the long-term. We could therefore deduce that, over the long-term, innovative companies perform less well. This conclusion seems logical, but it would be wrong. In fact, the opposite is true: innovative companies (assimilated here to growth stocks) perform "too well" over certain periods (as in our study period?). In fact, they perform so well that they reach valuations that, subsequently, strongly reduce their stock market potential.

Therefore, the long-term performance of these securities should not be confused with the stock market performance of the investors who buy these securities. Indeed, in the growth category, we find many innovative companies that are in a way "victims of their success". These companies are recognized by investors and are sometimes bought beyond their intrinsic value. First-time investors are therefore the ones who will make the biggest profits. Subsequently, in view of this high valuation, this limits the profit potential compared to so-called value companies which, sometimes bordering on default, have a much higher potential (as well as more risk).

The FF's three-factor model has therefore made it possible for us to better understand the characteristics and performance of innovative companies. However, summarizing innovative companies as growth companies seems closer to reality, but still a bit limiting. In fact, there are many different innovative companies. In our analysis, some of them appear as value companies. We have also limited our analysis to companies with larger market capitalization. We will continue our use of financial asset pricing models to better understand innovative companies by using the Carhart model.

6.1.4. *Carhart model*

The FF three-factor model was used as a reference for many years after its publication in 1993 in terms of its empirical efficiency. However, certain anomalies remain unexplained, including one in particular: the momentum effect.

The idea is to invest in stocks that have performed best historically (over the last 6–12 months), and hold these stocks for the next 6–12 months. This simple "technique" has historically produced a performance well beyond what one would expect based on the risk of the assets. It therefore tends to show an under-reaction of the markets, which would take time to incorporate the new information. This goes against the basic principle of market efficiency[10] and can only be explained by a bias in investor behavior, as highlighted by the followers of behavioral finance.

To account for this anomaly, Mark Carhart added a momentum factor to the Fama and French three-factor model.

Carhart's model is:

$$R_{it} - RFR_t = a_i + b_{i1}.(R_{mt} - RFR_t) + b_{i2}.SMB_t + b_{i3}.HML_t + b_{i4}.UMD_t + e_{it}$$

with UMD: difference between the returns of diversified up and down equity portfolios. Up shares are the best performers over the last 12 months and down shares are the losers over the last 12 months.

EXAMPLE.– Developing the Carhart model using Excel.

To analyze stock performance using a Carhart model, we must therefore add the momentum factor to the Fama and French model:

=INDEX(LINEST(excess return of the share to be analyzed; return of the 4 Carhart's factors;;TRUE);1;1)

Based on the above equation, ";1;1" will represent the coefficient of UMD, i.e. momentum. If the momentum is positive ("UP"), the coefficient

10 Market efficiency assumes that all new information is immediately incorporated by the markets and it is, therefore, not possible to generate excess (risk-adjusted) returns on the basis of this information.

will be positive and statistically significant. This means that the stocks that have performed best over the last 12 months are those that perform best again with an effect that could be described as "sheep-like". This should not happen, since markets are (normally) efficient, all new information should be immediately reflected in stock prices, which should adjust much faster to the good news. Unless investors are not (perfectly) rational, this should not happen so consistently. If, on the other hand, stocks that underperformed one year underperform again the next year, and do so consistently, this will be referred to as negative down momentum.

Be careful not to forget to select the four factors simultaneously when writing the equation in Excel.

	Alpha Carhart model, annualized	**Momentum**
Netflix	22.72%	/
Tesla	17.93%	/
US T-Mobile	11.87%	/
Amazon	8.54%	Up (*)
Boeing	7.75%	/
Facebook	6.44%	/
Microsoft	4.40%	Up (*)
Marriott	4.00%	/
JPMorgan Chase	3.67%	Up (*)
McDonald's	1.09%	Up (*)
Alphabet/Google	0.90%	Up (*)
Apple	0.54%	/
Salesforce	–0.42%	/
Johnson & Johnson	–0.72%	/
Cisco Systems	–3.41%	Down (*)
3M	–5.25%	/
Pfizer	–6.26%	/
AT&T	–8.30%	/
General Motors	–8.78%	/
IBM	–17.08% (*)	Down (*)

(*) Significant at a 5% threshold.

Table 6.4. *Ranking based on annualized apha of the 20 most innovative US companies and their momentum calculated using Carhart's four-factor model (source: Bloomberg and BCG (2019); Kenneth French website)*

Carhart thus added a factor to the FF model to develop a four-factor model that became a reference for many years despite the academic unease around this momentum factor.

It is interesting to note that, based on our sample, the performance of innovative companies does not seem to be mainly driven by a momentum effect. In fact, only seven companies out of 20 show a significant sensitivity to this factor. Amazon, Microsoft, JP Morgan Chase, McDonald's and Alphabet/Google all show positive momentum. We would therefore observe a "sheep-like" effect on these stocks, with investors choosing them in part because they "perform well". On the other hand, Cisco Systems and IBM have negative momentum! All the other companies show an insignificant sensitivity to this factor.

The momentum factor, although present, does not seem to be specific to innovative shares. We will therefore continue our review of financial asset valuation models and move on to the use of one of the most recent models, the Fama and French five-factor model.

6.1.5. *Fama and French five-factor model*

In 2015, Fama and French proposed a new version of their model by adding two factors: profitability and investments. It appears that companies with higher profitability tend to outperform and companies that invest the most (relative to their assets) tend to underperform. This approach should allow us to get rid of the momentum factor, which is quite embarrassing for the academic world.

This model should enable us to analyze the performance of innovative shares in greater detail:

$$R_{it} - RFR_t = a_i + b_{i1}.(R_{mt} - RFR_t) + b_{i2}.SMB_t + b_{i3}.HML_t + b_{i4}.RMW_t + b_{i5}.CMA_t + e_{it}$$

with:

– RMW: difference between the returns of diversified portfolios with relatively high (robust) and low (weak) profitabilities;

– CMA: difference between the returns of diversified portfolios with low (conservative) and high (aggressive) investments.

EXAMPLE.– Developing the Fama and French five-factor model using Excel.

To analyze stock performance using a Fama and French five-factor model, we need to add the profitability and investment factors to the initial Fama and French three-factor model. We should bear in mind that the three initial factors (RMRF, SMB and HML) are not perfectly identical to the initial model. It is advisable to download the five factors together from Kenneth French's site, since they are processed together to avoid correlation problems between them (multicollinearity):

=INDEX(LINEST(excess return of the share to be analyzed;return of the 5 FF factors;;TRUE);1;1)

Based on the above equation, ";1;1" will represent the CMA coefficient, i.e. the investment factor. Investment is measured by the growth of the company's assets.

If the coefficient of the investment factor is positive and statistically significant, the company will appear conservative. If this growth is consistently above-average, the CMA coefficient will appear negative and significant. The company will be classified as aggressive.

The ";1;2" factor is RMW, the profitability factor. Companies with consistently above-average operating profitability are qualified as robust. Conversely, they are weak.

In their 2015 article, Fama and French highlighted the fact that their model well explained the return of the largest market capitalizations. Our results confirm this observation. Indeed, we do not observe alpha (return not explained by risk) either, with the exception of IBM which seems to have been penalized beyond its risk during the period under review.

We also observe that newer innovative companies tend to be more "aggressive" in terms of investment. More mature innovative companies, on the other hand, exhibit more profitability and relatively less investments.

Note that Fama and French observed a different behavior between small and large market capitalizations. Thus, small market capitalizations with very high investments coupled with low profitability show very low returns, much lower than the model predicts.

It therefore seems that this type of company is particularly overvalued, which explains its subsequent stock market underperformance. This seems to confirm investors' attraction for innovative companies, apparently beyond the rational.

	Alpha five Fama and French factors, annualized	Profitability	Investments
Netflix	24.61%	/	Aggressive (*)
Tesla	17.58%	/	/
US T-Mobile	12.36%	/	/
Amazon	10.27%	/	Aggressive (*)
Facebook	8.77%	Weak (*)	Aggressive (*)
Boeing	6.99%	/	/
Microsoft	5.49%	Robust (*)	Aggressive (*)
JPMorgan Chase	5.28%	Weak (*)	Aggressive (*)
Alphabet/Google	3.11%	/	Aggressive (*)
Marriott	2.58%	Robust (*)	/
Salesforce	1.34%	Weak (*)	Aggressive (*)
Apple	0.94%	Robust (*)	Aggressive (*)
McDonald's	0.83%	Robust (*)	Conservative (*)
Johnson & Johnson	– 0.57%	/	Conservative (*)
Cisco Systems	– 3.26%	/	/
3M	– 5.76%	Robust (*)	/
Pfizer	– 5.89%	Weak (*)	Conservative (*)
AT&T	– 9.12%	Robust (*)	Conservative (*)
General Motors	– 9.85%	Robust (*)	/
IBM	– 18.31% (*)	Robust (*)	Conservative (*)

(*) Significant at a 5% threshold.

Table 6.5. *Ranking based on annualized alpha of the 20 most innovative U.S. companies and their sensitivity to the profitability and investment factors calculated using the FF five-factor model (source: Bloomberg and BCG (2019); Kenneth French website)*

6.1.6. *Sensitivity of innovative companies to risk factors*

In our analysis, we used the most commonly used financial asset valuation models. They offer the advantage of simplicity, and all our analyses are easily replicable with Microsoft Excel. The data is also easily accessible on sites like Yahoo Finance.

The question was to analyze whether innovative companies perform better or worse than the market and what their characteristics are. Our answer is nuanced.

There is no one kind of innovative company. Is it a young company or, on the contrary, a mature company? What is its sector of activity? The characteristics of the company can generate significantly different results.

Nevertheless, we can extract some findings from our results.

Newer innovative companies tend to be riskier than the market (high volatility, beta greater than 1) and have characteristics of growth companies. In addition, they are generally aggressive in their investments.

More mature innovative companies tend to be less risky than the market and have relatively more conservative investment characteristics and more robust profitability (the two are linked).

During the period under review, growth companies were strongly favored by the markets and the newer companies thus performed (much) better than the market overall. However, when adjusted for risk, none of these companies recorded an alpha statistically different from 0; so, it could be concluded that these companies outperformed but are riskier.

6.2. Outperformance of innovative companies

To develop our analysis, we will review the academic literature on the subject, in order to determine whether innovative firms are well valued by markets (based on the models described above).

As a reminder, if these companies are well evaluated, the constant of the regression or the "alpha" should be insignificant. If it is positively significant, it would tend to show that there is an under-reaction of the markets. If it is negatively significant, it would illustrate an over-reaction of the markets.

Hirshleifer *et al.* (2013) showed that the most innovative companies delivered significantly positive alpha. They used an original methodology based on the concept of "innovative efficiency". They showed that this effect was greater for small market capitalizations. They also observed a positive correlation between companies that had innovative efficiency and the profitability factor used in the FF five-factor model.

Houe *et al.* (2015) showed that companies with higher R&D intensity generated statistically significant alphas when using a model composed of size, profitability and investment factor. On the other hand, based on Carhart's model, which includes the momentum factor, the alpha is no longer significant. This would therefore tend to show that innovative companies are recognized and sought after by investors, sometimes beyond a rational valuation.

Hirshleifer *et al.* (2017) moved from their concept of innovative efficiency to a new concept of "innovative originality". Their study shows that firms with high innovative originality exhibit statistically significant future alphas. These results can be observed for both small and large market capitalizations. Innovative originality thus appears to be a precursor indicator of company profitability. It would therefore be possible to take advantage of the market's under-reaction to these innovative companies, which are still unknown.

These various articles therefore tend to show an abnormally positive return (alpha) for stocks with strong innovation, and explain this high performance by an under-valuation of the market, which is slow to recognize these innovations and values too weakly these companies that will subsequently perform more strongly.

7

Innovation and Credit Risk

For investors, the two main asset classes are equities and bonds. Equity markets were discussed in Chapter 6 and markets appeared to value, globally, innovation. This means that the stock prices of the most innovative companies, which manage to convert this into profit, tend to rise. We will now discuss the link between innovation and bond markets and, more specifically, credit risk.

7.1. Credit risk

In the bond markets, investors face two main risks: interest rate risk and credit risk.

Interest rate risk corresponds to the sensitivity of a bond to a rise (or fall) in interest rates. The idea is that if the general level of interest rates rises, this will result in a fall in bond prices. To understand this relationship, it is necessary to look at the general characteristics of bonds. The latter (in their classic formula, called plain vanilla) are defined by the following general characteristics:

– a final maturity: unlike equities whose "final maturity" is uncertain, at best infinite, bonds are defined by a final maturity. This is why we speak of investment with a fixed maturity;

– a fixed coupon rate: the coupon rate is the yield that will be paid each year to the investor by the issuer of the bond. The currency amount is calculated as a percentage of the nominal amount invested.

There are also some special features that are important for a good understanding of the product:

– the price of a bond is determined as a percentage of the nominal value;

– the maturity price of a conventional bond is 100, which means that at maturity, the issuer of the bond will reimburse you 100% of the face value.

EXAMPLE.– Calculation of the price of a bond.

Mr. Martin wishes to purchase 1 million euros worth of bonds from the oil company Total. The market price is 90. How much will Mr. Martin have to pay?

Since the price is calculated as a percentage of the nominal, Mr. Martin would have to pay 90% x 1,000,000 = €900,000. Conversely, if the price had been 110, he would have had to pay €1,100,000. This assumes that there is no accrued interest.

If a bond is purchased outside of the coupon payment date, the purchaser will be required to pay "accrued interest". This is the portion of the coupon that has "matured" since the last payment. For example, if the coupon payment date is June 30, and you buy this bond on December 31, this means that there are six months of coupon to be paid to the seller. This is justified since you will receive a full coupon on June 30 of the following year, even if you have only held the security for six months.

By clarifying these basic concepts, we can better understand the concept of a bond's yield to maturity.

A bond's yield to maturity is expressed on an annual basis and is determined by three main elements: the coupon rate, the capital gain or loss, and the credit risk (probability of default).

In the case of a risk-free bond (German government bonds are generally the benchmark in Europe), the YTM will therefore depend on the coupon rate and the capital gain and/or loss.

EXAMPLE.– Calculation of the yield to maturity of a bond (YTM).

Mrs. Smith buys a German government bond, maturing in 10 years, with a 1% coupon and at a price of 110. What is this bonds' yield to maturity?

This bond's yield to maturity will be close to 0%. In fact, we will "lose" 10% over 10 years on the price (we buy at 110 a security that will be worth 100 at maturity), i.e. around 1% per year. On the other hand, we will gain 1% coupon each year, 1% capital loss + 1% coupon = 0%.

This calculation is only an approximation, but it does give a good idea of the order of magnitude of the yield to be expected on a bond.

To be exact, the following formula must be applied to determine a bond's yield to maturity:

$$P = \sum_{t=1}^{T} \frac{CF_t}{(1+YTM)^t} = \frac{CF_1}{(1+YTM)^1} + \frac{CF_2}{(1+YTM)^2} + \cdots + \frac{CF_T}{(1+YTM)^T}$$

To accurately calculate the YTM, a financial calculator or a tool like Microsoft Excel is required. But the approximation described above based on the coupon and the price will give you an idea of the size.

For bonds that are not risk-free, a risk premium must be added. This is a yield supplement that investors will demand to cover the risk of capital loss. The higher the risk of loss, the higher the additional return will be. This risk will be reflected either in a higher coupon rate at issue or in a lower price if the risk deterioration has occurred after the issue.

This risk is called credit risk.

EXAMPLE.– Change in the price of a bond following a change in its credit risk.

Mr. Smith purchased bonds from Air France KLM at a price of 100. The annual coupon is 2% and the final maturity is in five years. In March 2020, the deterioration in the health situation following the COVID-19 crisis grounded the majority of the company's aircraft. What was the impact on credit risk, given that prices rose from 100 to 82 between February 19, 2020 and April 30, 2020?

This translates into an approximate increase (the exact YTM is given by the formula above) of the yield to maturity from 2 to 5.6%. The 18% decrease in price means that, if there is no default by the company, the investor will be repaid at the price of 100 at maturity. He will therefore "earn"

18% in five years, i.e. 3.6% per year. This 3.6% is in addition to the 2% coupon. The YTM has therefore gone from 2% to 5.6%. This reflects the increased risk of default perceived by investors who are now demanding greater compensation for accepting to take the risk on the company.

7.2. Credit risk and rating agencies

While the main objective of the investor in equity markets is to maximize the risk-adjusted return, what is important in the bond markets is to limit the risk of default as much as possible. The long-term bond investor mainly seeks to recover his or her capital at the end.

S&P	Moody's	Fitch
	Investment Grade	
AAA	AAA	AAA
AA+	AA1	AA+
AA	AA2	AA
AA–	AA3	AA–
A+	A1	A+
A	A2	A
A–	A3	A–
BBB+	BAA1	BBB+
BBB	BAA2	BBB
BBB–	BAA3	BBB–
	Speculative grade/High yield	
BB+	BA1	BB+
BB	BA2	BB
BB–	BA3	BB–
B+	B1	B+
B	B2	B
B–	B3	B–
CCC+	CAA	/
CCC	CA	CCC
CCC–	C	/
D	/	D

Table 7.1. *Rating tables of the main rating agencies (sources: Standard & Poor's, Moody's, Fitch)*

Analyzing the solvency of a company is not an easy exercise for individual investors. The expertise needed to carry out the analysis is already important, but more than a matter of expertise, it is a matter of time. Indeed, in order to be able to compare investments with each other, it is necessary to know their risk. A solvency analysis should therefore be carried out for all potential investments, which is not materially possible.

It is in this context that rating companies, the main ones being Moody's, Standard & Poor's and Fitch, have become indispensable.

They carry out this exercise of analyzing most of the companies in the world and provide us with a quick basis for comparison.

They thus establish the rating, of all these companies. The latter thus receive a rating ranging from AAA, the best quality, to D (the company is in default) (see Table 7.1). The investor will thus demand a return that is all the higher the greater the risk taken.

The question here is: what is the impact of innovation on corporate credit risk?

7.3. Relationship between credit risk and innovation

The idea is therefore to determine whether innovative companies are more or less at risk of default than those that are not.

Simply put, a company's solvency depends on two main elements: its relative debt ratio (and its ability to repay that debt) and its profitability. A company that is highly indebted in relation to its assets thus presents a non-negligible risk of no longer being able to repay its debt in the event of a problem. The more sensitive the company is to external factors, like the economic cycle, the more it will be perceived as risky and the lower the rating will be. On the other hand, the more profits the company generates, the more it will be able to honor its debt payments, which reduces the risk.

Innovation requires investment, and funding it puts the company at risk. Indeed, it is not at all certain that these investments translate into success and profits. The risk may come from technical elements, such as the green light from the competent authorities to launch a product, or from commercial elements. The product may be *a priori* excellent, but never meet its public.

Investments in innovation have an additional risk character compared to investment in assets due to their intangible nature. As Bloom (2007) showed in particular, if investments linked to an innovation are stopped, it is difficult to resell them or redeploy them to something else. The loss is almost total.

On the other hand, it is innovation, and the marketing of products that generate profits, which will make it possible to generate sufficient liquidity to finance the debt. Failure to innovate can also lead a company to become obsolete and lose market share. The example of telephony is very illustrative in this regard. The inability of Nokia and Ericsson to reinvent themselves with the arrival of Apple's iPhone brought these companies close to bankruptcy in just a few years.

It is therefore a question of finding the right balance between investments, and their financing, and the return on these investments[1].

Rating agencies, like Standard & Poor's (S&P Global Ratings 2013), include corporate profitability in the establishment of their ratings. This is how they do it; a company's ability to innovate and defend its competitive position is a central element in assessing a company's long-term solvency.

In the short-term, investments in innovation tend to have a negative value for bondholders. Indeed, the investment will result in a decrease in liquidity and/or an increase in the company's debt, all for a result that is uncertain. On the other hand, in the long-term, the development or even the survival of the company certainly depends on these investments.

As Chan *et al.* (2001) showed, if the value of intangible assets linked to investments in research and development is not taken into account, the firm's assets decrease and therefore the debt burden appears relatively higher, which will result in a lower rating and a higher yield to maturity. According to the accounting method adopted, the company may go so far as to exclude the value of its R&D expenses, which will lead to an underestimation of the company's book value, and therefore to an overestimation of its risk of default. This tends to be particularly true for companies with small market capitalizations that have less liquidity to face shocks. According to an article published by Franzen *et al.* (2007), incorporating the value of these R&D expenditures would make it possible

1 ROI.

to reduce the probability of default and raise the rating of these companies. This overestimation of their risk also certainly allows for a better understanding of the outperformance of smaller-cap stocks over the long-term, which tend to underperform what the (unadjusted) models predict.

This underestimation of the potential of R&D spending by companies and the consequences in terms of innovation may also explain why managers of so-called "insiders" tend to achieve excess returns compared to the market, even when all official publication rules are respected. This is what is known as the "inefficiency of strong form markets", as defined by Eugène Fama in 1969.

7.4. Innovative companies less risky in the long-term?

The literature on the subject shows mixed results. The difficulty of this exercise is defining the "right" measure for innovation. Spending on R&D sometimes leads to a weakening of the company's balance sheet structure and an increase in its risk. This is illustrated by Shi's study (2003). Zhang (2015) confirmed in an article the negative relationship between R&D spending and credit quality. They showed that companies with the weakest balance sheet quality are the most exposed.

On the other hand, in an article, Hsu *et al.* (2015) showed that a more refined analysis of the problem, which takes into account the performance of innovation (originality of patents, impact, number), tends to reduce the risk perceived by investors in the most innovative companies. Eberhart *et al.* (2008) also concluded that there was a positive relationship between innovation and credit quality by measuring the "intensity" of R&D.

Ciftci and Darrough (2016) looked at so-called "private" debts, i.e. debts that do not have an official rating. It is generally the smallest companies that use this type of investment. These results are therefore important because they extend the conclusions to this market segment, which includes many innovative companies. Their results show a positive relationship between R&D spending and credit quality.

Another way to study innovation is to analyze the number of patents that are filed. The increase in the number can either illustrate a stronger innovation or simply a stronger communication about these innovations. In the latter case, it reduces the asymmetry of information between companies and investors and leads to an increase in the value of companies. Czarnitzki and Kraft (2004) showed a positive relationship between the number of patents and credit quality. Liao (2013) confirmed the previous results based on the number, value and income related to patents. Hsu *et al.* (2015) and Griffin *et al.* (2016) tended to confirm these results also on the basis of slightly different measures. Thus, it seems that patents, and their quality, are well incorporated in the calculation of a firm's rating and, consequently, on the credit spread.

An additional question is whether the most innovative companies are more subject to a default risk. Indeed, this race for innovation, if not won, can be very costly. The example of Nokia in recent years is very illustrative in this respect.

Eisdorfer and Hsu (2011) showed that the degree of technological watch in a sector (measured by the number of patents filed) is correlated with the number of patents filed for bankruptcies. Companies whose innovative performance is relatively weak are thus more affected in highly innovative sectors. The authors showed that classical default prediction models can be improved by taking into account the innovative capacity of companies in these highly innovative sectors.

7.5. Conclusion

The link between credit risk and business innovation seems to be well established. On the one hand, more innovative companies, which spend more on research and development, appear to be more at risk. Indeed, these companies are often active in very competitive and constantly evolving sectors. When these investments are not converted into profit, the consequences can be dramatic.

But it also highlights the importance of taking into account a finer variable, which is the ability of companies to transform these investments into success, namely innovative performance.

Taking these elements into account should enable credit analysts to better understand the risk of default and therefore better estimate the rating of companies.

8

Financial Innovation: Creating Value… or Risk?

8.1. Basic concepts of portfolio management

In this chapter, we will try to answer, based on the fundamental concepts of portfolio management, the following question: does financial innovation create value or risk for the investor? In order to do so, we will first go back to the fundamental concepts of portfolio management.

8.1.1. *Diversification*

The principle of diversification is widely known, both by financial asset management professionals and private individuals. However, while it seems simple, it is often misunderstood.

The general idea is that by combining different assets, we will reduce the risk of our portfolio. That's right, but we need to be more specific!

In fact, it is important to return to the basic objective: what is the purpose of diversification? Yes, the objective is to reduce the risk of our portfolio, but in a very specific context: when the markets fall and in an extreme case, in the event of a crash.

Indeed, it is when the markets fall that investors want this decline to be cushioned. But cushioned by what? If you have a large number of financial assets in the portfolio, but they all fall at the same time, this is not very useful. This last remark may seem rather obvious, but it is very important to remember, as markets sometimes behave in a way that is described as gregarious. Thus, when equity markets fall, many financial assets suffer the same fate: real estate, commodities, high-yield bonds. This is typically what was observed during the financial crisis of 2008 and again during the recent COVID-19 crisis.

Thus, there are many assets in the markets that tend to move "increasingly" in the same direction when markets fall. This is called correlation, which increases to tend towards 1, i.e. 100% correlation. There is then no more diversification in this case.

On the other hand, there are assets that have the opposite properties: when equity markets fall, these assets tend to rise; investors favor these assets when the context is particularly uncertain. We are talking here about so-called "risk-free" bonds, such as German or American government bonds, gold, or other types of assets perceived as safe havens.

8.1.2. *Markowitz model*

It was as early as 1952 that Professor Harry Markowitz specified and quantified the central concept of diversification and, as an extension, portfolio management.

Again, our goal here is not to get into theory, but rather into the advantage and practical application of the concepts. Readers wishing to go further into these elements can refer to the bibliography.

Markowitz laid the foundation for building a portfolio of financial assets. To do so, he summarized the problem by determining three basic inputs: expected return, expected risk and correlations between assets. This last ingredient is in a way the magic ingredient of portfolio management.

Thus, for the same level of expected return, for example, reaching 5% return over the year, it is possible to reduce risk by combining assets that are not perfectly correlated with each other. The expected return is therefore the same, but the risk of loss is reduced.

This is why diversification is central to portfolio management.

8.1.2.1. *Expected return on a portfolio of financial assets*

Calculating the expected return on a portfolio of financial assets is fairly simple; it is the weighted average of the expected returns on the assets that make up the portfolio.

EXAMPLE.– Calculation of the expected return on a portfolio.

Ms. Smith has a portfolio of two financial assets: stocks and bonds. She owns 25% in stocks and 75% in bonds. Knowing that the expected return on equities is 10% and that the return on bonds is 4%, what is the expected return on her portfolio?

It's very simple; the expected return on a portfolio is simply the weighted average of the individual returns. In this case, it is therefore:

(25% x 10%) + (75% x 4%) = 2.5% + 3% = 5.5%.

Diversification is therefore not, as is often mistakenly thought, a question of return. You can increase the number of financial assets in your portfolio as much as you want, the return on the portfolio will always be the weighted average of the different returns.

8.1.2.2. *Expected risk of a portfolio of financial assets*

Diversification is therefore exclusively a question of risk in the Markowitz model.

Portfolio risk, when assets are not perfectly correlated, is lower than the weighted average risk of the individual securities. Thus, the more diversified the portfolio, the lower the risk.

To measure risk in the Markowitz framework, the concept of standard deviation is used:

$$\sqrt{\frac{1}{n}\sum_{i}\left(x_i-\overline{x}\right)^2}$$

Behind this formula lies the following idea: by how much do the observed returns deviate from the average returns?

So, if your goal is 5%, it is not the same thing if you get it with a smooth linear increase, or if you get it with an average of returns ranging from –20% to +20%.

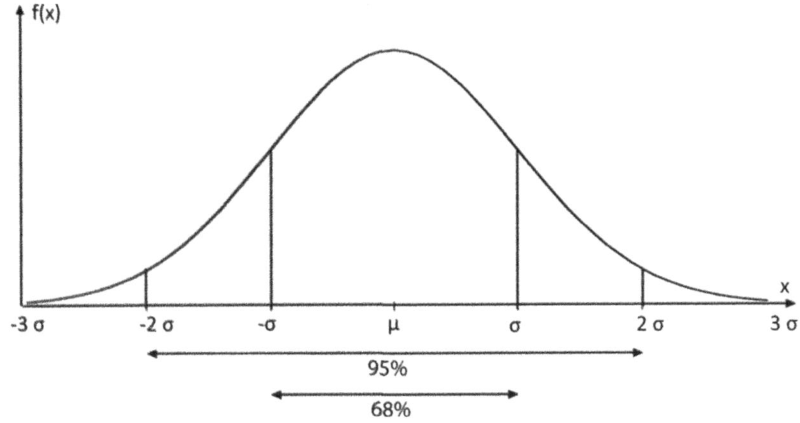

Figure 8.1. *Normal distribution*

Thus, the higher the standard deviation, the greater the risk. If the standard deviation is 20% (which is close to the historical average for equities), this means that there is a 68% chance of being in the range: mean – 20% to mean + 20%. There is a 95% to be between mean – 40% to mean + 40%.

Portfolio risk is equal to the weighted average of standard deviations minus the benefits of diversification. Thus, the portfolio risk will only be equal to the weighted average of the individual risks in the case of a perfect correlation between these assets.

8.1.2.3. *Covariance and correlation concepts*

The tool to measure diversification is covariance. As you can see below, it is a variant of variance. The idea is to cross-tabulate the different assets with each other.

The lower the covariance, the greater the reduction in portfolio risk.

The portfolio risk is determined by the following formula:

$$\sigma_{port} = \sqrt{\sum_{i=1}^{n} w_i^2 \sigma_i^2 + \sum_{i=1}^{n} \sum_{i=1}^{n} w_i w_j Cov_{ij}}$$

where:

– σ_{port} = standard deviation of the portfolio;

– W_i = the weights of the individual assets in the portfolio, with these weights being determined by their proportion of value in the portfolio;

– σ_i^2 = variance of the rate of return of asset *i*;

– Cov_{ij} = covariance between the rate of return of assets *i* and *j*, where $Cov_{ij} = r_{ij}\sigma_i\sigma_j$.

It is well observed that the lower the covariance, the lower the risk of the portfolio.

For an easier reading of this concept, correlations are generally favored:

$$r_{ij} = \frac{Cov_{ij}}{\sigma_i \sigma_j}$$

where:

– r_{ij} = correlation coefficient of returns;

– σ_i = standard deviation of R_{it};

– σ_j = standard deviation of R_{jt}.

It is in fact a standardized version of covariance by dividing the covariance by the product of the standard deviations. This has the advantage of ranging correlations between – 1 and + 1.

A correlation of +1 means a perfect correlation and therefore no diversification benefits. The ideal is to obtain the lowest possible correlation, and even ideally a decorrelation or negative correlation. Thus, in the event of a fall in one asset, this fall is compensated by the rise in another asset. In the event of a financial crisis, this is generally what is observed between equities and bonds. Investors sell their shares in these difficult economic contexts in favor of assets perceived as safe, such as government bonds. As a result, there is a decline in equities and a rise in bond prices. The correlation is thus negative.

8.1.2.4. *Markowitz's optimal portfolio*

On the basis of the different financial assets available, it is therefore possible to combine them. It appears that it is possible to combine them in such a way that, for a certain level of expected return, there is only one "optimal" combination of these assets that minimizes the portfolio risk. No other combination can achieve a lower risk for this objective. Conversely, for a given level of risk, there is only one combination that gives the highest expected return.

The combination of these different optimal portfolios makes up the efficient Markowitz frontier.

EXAMPLE.– Determining the efficient frontier using Excel.

In practice, it is relatively easy to determine the efficient frontier using Microsoft Excel. To do so, we need to collect the expected returns, as well as the variance–covariance matrix. By using this historical matrix, we implicitly assume its stability.

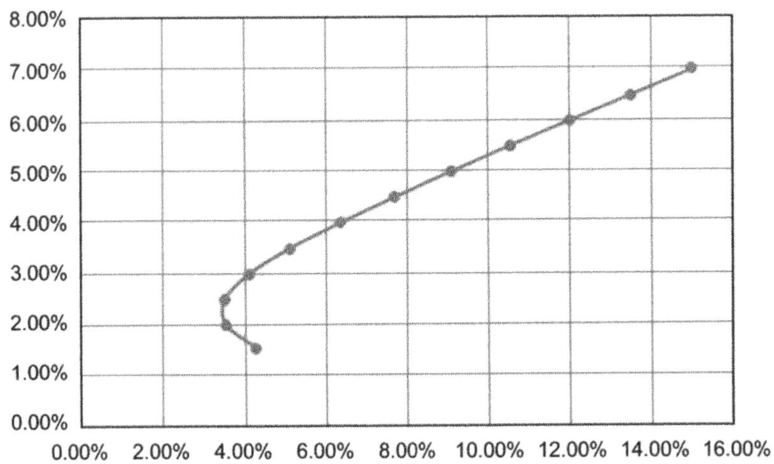

Figure 8.2. *Example of an efficient frontier calculated using Excel (source: Bloomberg). For a color version of this figure, see www.iste.co.uk/dispas/innovation.zip*

In Figure 8.2, we constructed the efficient frontier of a portfolio with six assets: U.S. equities represented by the S&P 500, U.S. government bonds, U.S. Corporate Investment Grade bonds, a commodities index (Goldman Sachs Index), a U.S. High Yield index (Bloomberg Barclays Index) and a U.S. real estate index (FTSE Index).

The Markowitz model is very useful for understanding and constructing a portfolio of financial assets, as it provides us with a pragmatic answer to the problem of portfolio construction; it tells us "how much" to invest in each asset (optimal weights) to build the portfolio, which is exactly what investors are looking for.

To determine these optimal weights, we therefore collected the following information: expected returns, expected risks and expected covariance (correlations). To do so, we used the financial asset valuation models described in Chapter 7. In the above example, we used the CAPM with a risk-free rate of 2% and a market risk premium of 5%. We calculated the betas over the period 1998–2019 on a monthly basis.

For the variance–covariance matrix, we first composed series of excess returns, i.e. return – risk-free rate. In Excel, on this basis, we reconstruct the variance–covariance matrix using the following formula:

= mmult(transpose(returns-risk free rate);returns- risk-free rate)/(number of data – 1)

For three years of monthly data, there are therefore 36 data - 1 = 35 data. To calculate this matrix, the following procedure should be followed:

1) select the final zone. This means that if we work on a portfolio of 20 stocks, the VCOV matrix will be 20 x 20. It is therefore advisable to select this area of 20 rows and 20 columns in Excel;

2) type the formula described above;

3) press the Control and Shift keys at the same time. Keep these two keys pressed and press Enter to get the final result.

The sequence explained must be followed for the results to appear.

VCOV matrix

0.001875157	-0.0001713	0.0001261	0.000707477	0.00087791	0.001488759
-0.00017127	0.00015389	0.00011262	-5.87293E -05	-0.0001191	-1.64466E -05
0.000126102	0.00011262	0.00022627	0.000204618	0.00012309	0.00033139
0.000877905	-5.873E -05	0.00020462	0.000663986	0.00057043	0.00093393
0.000877905	-0.0001191	0.00012309	0.000570427	0.00426352	0.000810453
0.001488759	-1.645E -05	0.00033139	0.00093393	0.00081045	0.003508779

Figure 8.3. *Example of the variance and covariance matrix calculated using Excel*

Then, using this VCOV matrix, it is possible to calculate the risk of the portfolio. To do this, the following formula must be applied in Excel: X' V X. X represents the weights of the assets. X' is the "transpose" of X. X is a column of 20 securities in the previous example. X' is therefore a row with these 20 securities.

The formula for calculating the variance of the portfolio is as follows: asset weight (row) × VCOV matrix × asset weight (column), which in Excel translates into:

=mmult(mmult(assets weights (rows); VCOV matrix);assets weights (columns)))

To obtain the result, press Control–Shift–Enter again simultaneously. To go from the variance of the portfolio to the standard deviation of the portfolio, the following formula must be applied (for monthly data):

=sqrt(portfolio variance) x sqrt(12)

On this basis, the two main elements are available: the expected return and the expected risk (adjusted for diversification *via* covariance).

8.1.3. *Financial innovation "theoretically" a source of diversification*

Insofar as financial innovations constitute new investment opportunities, the correlation of these new asset classes with traditional assets will be less than 1, and therefore there will be a contribution to portfolio diversification.

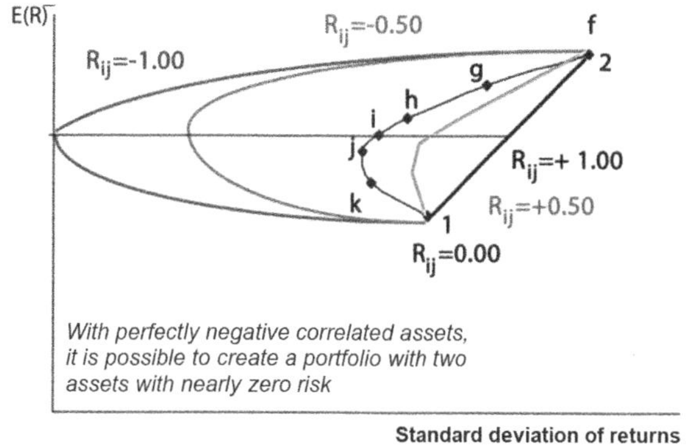

Figure 8.4. *Efficient frontiers as a function of asset correlation (inspired by Reilly and Brown (2012)). For a color version of this figure, see www.iste.co.uk/dispas/innovation.zip*

8.2. Risk-creating financial innovation?

Unfortunately, in reality, these new share classes tend to correlate when markets are in crisis. For example, for investors who had favored private equity, commodities, real estate, hedge funds, high-yield bonds, etc., these new classes of shares tended to be highly correlated during the financial crisis of 2008. As a result, the benefits of diversification were not generated when the investor needed it the most.

	1998–2007	2008–2019
U.S. Government Bonds	– 33.18%	– 30.69%
U.S. Inv. Corporate Bonds U.S. Rank	– 3.82%	34.13%
Raw materials index	– 1.74%	57.31%
High yield US	50.18%	72.28%
Real estate American side	32.27%	73.19%

Table 8.1. *Correlations between the S&P 500 and various so-called "alternative" asset classes (source: Bloomberg; Blackstone (2019))*

Following the example of Blackstone (2019), we have calculated the correlation of different classes of so-called "alternative" shares with respect to the S&P 500 American stock market index. It appears that the correlations have tended to increase between 2008 and 2019 compared to 1998 and 2007. For example, the correlation between listed real estate and the S&P 500 rose from 0.3 to 0.7. The correlation with commodities increased from 0 to 0.5.

Therefore, it appears that investors have globally faced an increase in the volatility of their portfolios, as they have replaced core assets such as listed stocks and (risk-free) bonds with assets whose risk they have underestimated. Collateralized Debt Obligations (CDOs) are the most obvious example with many investors substituting conventional bonds for this type of toxic product.

While financial theory tells us that financial innovations should promote diversification and thus reduce risk, it appears that the reality may be different. As illustrated in Table 8.1, two of the most uncorrelated assets are the basic historical assets: equities and government bonds. The problem is that if diversification is done at the expense of these assets, the result may be a riskier portfolio. For example, if we replace government stocks with high-yield bonds, commodities or real estate, we can see that the effect will be an increase in correlations and portfolio risk, not the other way around.

Therefore, asset multiplication should not be confused with diversification. Diversification is the consequence of the combination of uncorrelated assets. The financial crisis has been a strong indicator of this problem. Indeed, many new assets have brought little diversification in times of crisis. Derivatives are, as the name suggests, linked or derived from core assets.

Academic literature tends to confirm this conclusion. Buss *et al.* (2018) showed in an article that when non-professional investors are confronted with new financial assets, which they are less familiar with, there is an increase in the volatility of their portfolios. Simsek (2013a, 2013b) highlighted the same result due to the poor valuation given to these new products by inexperienced investors who risk investing in spite of too high a valuation. Moreover, these new types of investments, such as derivatives, private equity, hedge funds, natural resources or cryptocurrencies, tend to increase the speculative nature of the portfolio and thus its risk. Ang and Sorensen (2013), for example, illustrated the difficulties related to the valuation of private equity. Blackstone (2019) highlighted the appetite of investors for alternative investments, the potential gains in terms of diversification or return, as well as the difficulty of evaluating them accurately.

Conclusion

The first edition of this book is published at a particular time in terms of innovation and financial market developments.

On the one hand, the financial markets experienced, in March 2020, their biggest daily fall in more than 30 years – after the famous crash of 1987. At the time of writing, with more than half of the world still under lockdown, the economic effects of the COVID-19 health crisis are not yet fully assimilated by our economies and perhaps by the financial markets. The economic recessions that are certain to come, directly as a result of this crisis, may in some countries be the worst since the Great Recession of the 1930s. It is possible that the financial markets that have partially recovered their losses of March 2020 may plunge back due to the health, human, and financial disasters that will result from the tsunami we are currently experiencing. The wave is far from having receded, and it is unfortunately only when the final ebb is over that we will take measure of the human and economic damages inflicted by this crisis.

On the other hand, the only lifeline to avoid another market plunge will come from the colossal R&D and innovation capabilities invested by governments, research labs, biotech and the world's major pharmaceutical companies.

This crisis also sadly illustrates some of the concepts discussed in this book: the need for companies to have a permanent capacity to reinvent themselves and innovate; otherwise, they risk disappearing in the event of a deep external shock.

Who could have imagined a few months ago that half of the world's population would be under lockdown?

The effects are disastrous in every respect for many sectors (such as transportation, tourism or hotels and restaurants, for example). Many companies – often small- and medium-sized enterprises (SMEs) that are usually in a tight cash flow situation – risk going bankrupt.

At the same time, there are opportunities in every adverse situation; companies offering solutions that facilitate teleworking are growing, many solutions that benefit the local economy are exploding, and there is no doubt that some organizations will be born out of this crisis.

In a world that seems stable, black swans (Taleb 2007) will continue to appear. Our world will regain some post-COVID-19 stability, but similar events will occur in the future. More than ever, investing in the resources necessary for innovation seems crucial for companies.

References

Abdelkafi, N., Makhotin, S., Posselt, T. (2013). Business model innovations for electric mobility: What can be learned from existing business model patterns? *International Journal of Innovation Management*, 17(1), 1–41.

Agence française pour le développement et la promotion de l'agriculture biologique (2018). Les chiffres 2018 du secteur Bio. Document, Agence BIO, Montreuil.

Andersen, B. and Striukova, L. (2010). Where value resides in the modern enterprise. *Strategic Change*, 19(3–4), 103–123.

Ang, A. and Sorensen, M. (2013). Investing in private equity. *Alternative Investment Analyst Review*, 2, 21–31.

Ardichvili, A., Cardozo, R., Ray, S. (2003). A theory of entrepreneurial opportunity identification and development. *Journal of Business Venturing*, 18(1), 105–123.

Banque de France (2018). Le financement des entreprises. Document, Banque de France, Paris.

Bieger, T. and Reinhold, S. (2011). Das wertbasierte geschäftsmodell – ein aktualisierter strukturierungsansatz. In *Innovative Geschäftsmodelle*, Bieger, T., Zu Knyphausen-Aufseß, D., Krys, C. (eds). Springer, Berlin.

Blackstone (2019). *Seeking an Alternative: Understanding and Allocating to Alternative Investments*. Blackstone Publication, New York.

Block, J.H., Colombo, M.G., Cumming, D.J., Vismara, S. (2017). New players in entrepreneurial finance and why they are there. *Small Business Economics*, 50, 239–250.

Bloom, N. (2007). Uncertainty and the dynamics of R&D. *American Economic Review*, 97, 250–255.

Bökberg, A., Carrellas, J., Chau, A. (2019). Private markets come of age. Report, McKinsey Global Private Markets Review, New York.

Bontis, N. (2001). Assessing knowledge assets: A review of the models used to measure intellectual capital. *International Journal of Management Reviews*, 3(1), 41–60.

Bruton, G.D. and Prasad, D. (1997). Strategy and IPO market selection: Implications for the entrepreneurial firm. *Journal of Small Business Management*, 35(4), 1–10.

Bussy, A., Uppal, R., Vilkov, G. (2018). Financial innovation and asset prices. Working Paper, CEPR, London.

Carhart, M. (1997). On persistence of mutual fund performance. *The Journal of Finance*, 52, 57–82.

Carnitzki, D. and Kraft, K. (2004). Innovation indicators and corporate credit ratings: Evidence from German firms. *Economics Letters*, 82, 377–384.

CB Insights (2018). Venture capital funnel shows odds of becoming a unicorn are about 1%. Document, CB Insights, New York.

Chan, L.K.C., Lakonishok, J., Sougiannis, J. (2001). The stock market valuation of research and development expenditures. *Journal of Finance*, 56, 2431–2456.

Chemmanur, T.J., Krishnan, K., Nandy, D.K. (2011). How does venture capital financing improve efficiency in private firms? A look beneath the surface. *The Review of Financial Studies*, 24(12), 4037–4090.

Chen, N.-F., Roll, R., Ross, S.A. (1986). Economic forces and the stock market. *Journal of Business*, 59(3), 383–403.

Chesbrough, H. (2007). Business model innovation: It's not just about technology anymore. *Strategy and Leadership*, 35(6), 12–17.

Chesbrough, H. (2010). Business model innovation: Opportunities and barriers. *Long Range Planning*, 43(2), 354–363.

Christensen, C.M. (1997). The innovator's dilemma – When new technologies cause great firms to fail. *Harvard Business Review Press*, Cambridge.

Ciftci, M. and Darrough, M. (2016). Does the riskiness of R&D outweigh its benefits? The perspective of US private lenders. *Journal of Business Finance and Accounting*, 43(5), 654–692.

Conley, J.G., Bican, P.M., Ernst, H. (2013). Value articulation: A framework for the strategic management of intellectual property. *California Management Review*, 55, 102–120.

Cooper, A., Gimeno-Gascon, J., Woo, C. (1994). Initial human and financial capital as predictors of new venture performance. *Journal of Business Venturing*, 9(5), 371–395.

Dahlqvist, J., Davidsson, P., Wiklund, J. (2000). Initial conditions as predictors of new venture performance: A replication and extension of the Cooper *et al.* study. *Enterprise and Innovation Management Studies*, 1(1), 1–17.

Datar and INSEE (2000). Le taux de création et de survie des entreprises. *Observatoire des territoires*.

De Clercq, D., Fried, V.H., Lehtonen, O., Sapienza, H.J. (2006). An entrepreneur's guide to the venture capital galaxy. *Academy of Management Perspectives*, 20(3), 90–112.

Delmar, F. (1994). The risk management of the entrepreneur: An economic-psychological perspective. *Journal of Enterprising Culture*, 2(2), 735–751.

Delmar, F. and Shane, S. (2004). Legitimating first: Organizing activities and the survival of new ventures. *Journal of Business Venturing*, 19, 385–410.

Dorf, R. and Byers, T. (2005). *Technology Ventures: From Idea to Enterprise.* McGraw Hill, New York.

Douglas, E.J., Carlsson-Wall, M., Hjelström, T. (2015). Negotiating equity share and management control of the entrepreneurial new venture. *Venture Capital*, 16(4), 287–307.

Drif, A. (2019). Le "Private Equity" mondial prêt à détrôner la Bourse. *Les Échos*, March.

Duttaa, S. and Foltab, T. (2016). A comparison of the effect of angels and venture capitalists on innovation and value creation. *Journal of Business Venturing*, 31.

Eberhard, A.C., Maxwell, W.F. and Siddique, A.R. (2008). A reexamination of the tradeoff between the future benefits and riskiness of R&D increases. *Journal of Accounting Research*, 46(1), 27–53.

Eisdorfer, A. and Hsu, P.-H. (2011). Innovate to survive: The effect of technology competition on corporate bankruptcy. *Financial Management*, 1087–1117.

Fama, E.F. (1969). Efficient capital markets: A review of theory and empirical work. *Journal of Finance*, 25(2), 383–417.

Fama, E.F. and French, K.R. (1992). The cross section of expected stock returns. *Journal of Finance*, 46, 427–466.

Fama, E.F. and French, K.R. (1993). Common risk factors in the returns on stocks and bonds. *Journal of Financial Economics*, 33, 3–56.

Fama, E.F. and French, K.R. (1995). Size and book-to-market factors in earnings and stock returns. *Journal of Finance*, 50, 131–155.

Fama, E.F. and French, K.R. (1996). Multifactor explanation of asset pricing anomalies. *Journal of Finance*, 51, 55–84.

Fama, E.F. and French, K.R. (1998). Value versus growth: The international evidence. *Journal of Finance*, 53, 1975–1999.

Fama, E.F. and French, K.R. (2015). A five factor asset pricing model. *Journal of Financial Economics*, 116(1), 1–22.

Fédération du commerce et de la distribution (2019). Évolutions du commerce et de la distribution : faits et chiffres [Online]. Available at: http://www.fcd.fr/qui-sommes-nous/actualites-de-la-fcd/detail/publication-du-livret-fcd-evolutions-du-commerce-et-de-la-distribution-faits-et-chiffres-2019/.

Franzen, L.A., Rogers, K.J., Simin, T.T. (2007). Measuring distress risk: The effect of R&D intensity. *Journal of Finance*, 42(6), 2931–2967.

Gaymard, C., De Blignieres, G., Marchal, O. (2015). Accompagner les jeunes entreprises de croissance. Étude sur l'accompagnement des jeunes entreprises de croissance. Document, Raise, Bain & Company, Paris.

Gong, Y., Baker, T., Miner, A. (2009). Organizational routines and capabilities in new ventures. Document, Babson College Entrepreneurship Research, Babson Park.

Gosh, (2012). Why most venture-backed companies fail. Harvard business school newsroom, 10 December [Online]. Available at: https://www.hbs.edu/news/Pages/item.aspx?num=214.

Greene, F.J. and Hopp, C. (2017). Are formal planners more likely to achieve new venture viability? A counterfactual model and analysis. *Strategic Entrepreneurship Journal*, (11)1, 36–60.

Griffin, P.A., Hong, A.H., Woo Ryou, J. (2016). Corporate innovative efficiency: Evidence of effects on credit ratings, *Journal of Corporate Finance*, 51(C).

Hellmann, T. (2007). Entrepreneurs and the process of obtaining resources. *Journal of Economics and Management Strategy*, 16(1), 81–109.

Hellmann, T. and Puri, M. (2002). Venture capital and the professionalization of start-up firms: Empirical evidence. *The Journal of Finance*, (57)1, 169–197.

Hirshleifer, D.A., Hsu, P.-H., Li, D. (2012). Innovative efficiency and stock returns. *Journal of Financial Economics*, 107, 632–654.

Hirshleifer, D.A., Hsu, P.-H., Li, D. (2017). Innovative originality, profitability, and stock returns. *Review of Financial Studies*, 31(7), 2553–2605.

Holm, A.B., Günzel, F., Ulhøi, J.P. (2013). Openness in innovation and business models: Lessons from the newspaper industry. *International Journal of Technology Management*, 61(3), 324–348.

Honig, B. (2004). Entrepreneurship education: Toward a model of contingency-based business planning. *Academy of Management Learning and Education*, 3(3), 258–273.

Hou, K., Xue, C., Zhang, L., (2015). Digesting anomalies: An investment approach. *Review of Financial Studies*, 28, 650–705.

Hsu, P.-H., Lee, H.-H., Liu, A.Z., Zhang, Z. (2015). Corporate innovation, default risk, and bond pricing. *Journal of Corporate Finance*, 35, 329–344.

Hurst, D.K. (1995). *Crisis and Renewal: Meeting the Challenge of Organizational Change*. Harvard Business School Press, Boston.

Iaquinto, A. and Spinelli, S. (2006). *Never Bet the Farm: How Entrepreneurs Take Risks, Make Decisions – And How You Can, Too*. Jossey-Bass, San Francisco.

Initiative France (2017). Rapport d'activité annuel. Report, Initiative France, Paris.

Kaplan, R.S. and Norton, D.P. (2004). *Strategy Maps – Converting Intangible Assets into Tangible Outcomes*. Harvard Business School Press, Boston.

Kaplan, S. and Stromberg, P. (2002). Financial contracting meets the real world: An empirical study of venture capital contracts. *Review of Economics*, 70(2), 281–315.

Kaplan, S. and Stromberg, P. (2009). Leveraged buyouts and private equity. *Journal of Economic Perspectives*, 23(1), 121–146.

Kerr, W.R., Lerner, J., Schoar, A. (2014). The consequences of entrepreneurial finance: Evidence from angel financings. *The Review of Financial Studies*, 27(1), 20–55.

KPMG (2007). Les PME qui grandissent, qui sont-elles ? Pourquoi sont-elles si performantes? Report, KPMG Entreprises.

Lakonishok, J., Shleifer, A., Vishny, R. (1994). Contrarian investment, extrapolation and risk. *Journal of Finance*, 49(5), 1541–1578.

Lerner, J., Sorensen, M., Strömberg, P. (2011). Private equity and long-run investment: The case of innovation. *The Journal of Finance*, 66(2), 445–477.

Levis, M. (2011). The performance of private equity-backed IPOs. *Financial Management*, 40(1), 253–277.

Levratto, N., Tessier, L., Zouikri, M. (2011). L'innovation fait-elle partie des déterminants de la croissance des entreprises ? Une analyse des données françaises entre 1997 et 2007. In *PME, dynamiques entrepreneuriales et innovation*, Hamdouche, A., Reboud, S., Tanguy, C. (eds). Peter Lang, Brussels.

Levratto, N., Tessier, L., Fonrouge, C. (2016). Business performance and angels presence: A fresh look from France 2008–2011. *Small Business Economics*, 50(2), 339–356.

Liao, C.-H. (2013). R&D performance and credit ratings. *Journal of Accounting, Finance, and Economics*, 3(2), 53–71.

Lin, C., Jiang, J.L., Wu, Y.J., Chang, C.C. (2011). Assessment of commercialization strategy using R&D capability. *Industrial Management and Data Systems*, 111(3), 341–69.

Ministère de l'économie et des finances et Ministère de l'agriculture (2017). Observatoire de la formation des prix et des marges des produits alimentaires. Report to Parliament, Paris.

Mitteness, C.R., Baucus, M.S., Norton Jr., W.I. (2013). Establishing cognitive legitimacy in emerging organizations: The role of prestige. *Journal of Small Business Strategy*, 23(1), 71–91.

O.B. (2018). 84 % des ETI surperformantes sont soutenues par le private equity. *Le Magazine des Affaires*, Neuilly sur seine.

Observatoire du financement des entreprises, Ministère des finances et des comptes publics, Ministère de l'économie, de l'industrie et du numérique (2015). Rapport sur le financement des PME et ETI en croissance. Report, Observatoire du financement des entreprises, Paris.

OECD (2005). *Manuel d'Oslo : principes directeurs pour le recueil et l'interprétation des données sur l'innovation*, 3rd edition. OECD, Paris.

Osterwalder, A. and Pigneur, Y. (2011). *Business model nouvelle génération : un guide pour visionnaires, révolutionnaires et challengers*. Pearson, Paris.

Osterwalder, A., Pigneur, Y., Tucci, C.L. (2005). Clarifying business models: Origins, present, and future of the concept. *Communications of the Association for Information Systems*, 16(1), 1–25.

Popov, A. and Roosenboom, P. (2009). Does private equity investment spur innovation? Evidence from Europe. Working Paper, European Central Bank, Frankfurt am Main.

Porter, M.E. and Kramer, M.R. (2006). Strategy and society: The link between competitive advantage and corporate social responsibility. *Harvard Business Review*, 84(12), 78–92, December.

Rayna, T. and Striukova, L. (2016a). 360° business model innovation: Toward an integrated view of business model innovation. *Research Technology Management*, 59(3), 21–28.

Rayna, T. and Striukova, L. (2016b). From rapid prototyping to home fabrication: How 3D printing is changing business model innovation. *Technological Forecasting and Social Change*, 102, 214–224.

Reilly, F. and Brown, K. (2012). *Investment Analysis and Portfolio Management*, 10th edition. South Western Cengage Learning, Boston.

Réseau Entreprendre (2019). Étude panorama. Study, Réseau Entreprendre.

Ries, E. (2011) *The Lean Start-up: How Today's Entrepreneurs Use Continuous Innovation to Create Radically Successful Businesses*. Crown Business, New York.

Ringel, M., Grassl, F., Baeza, R., Kennedy, D., Manly, J. (2019). The most innovative companies 2019: The rise of AI, Platforms, and Ecosystems. Document, Boston Consulting Group, Boston.

Rogers, E. and Paul, J. (2018). Strategic people practice in start-up organizations. *People + Strategy*, 41(3), 32–36.

S&P Global Ratings (2013). Corporate methodology [Online]. Available at: https://www.spratings.com/scenario-builder-portlet/pdfs/CorporateMethodology.pdf.

Sahlman, W.A. (1997). How to write a great business plan. *Harvard Business Review*, July–August, 98–108.

dos Santos Paulino, V. and Tahri, N. (2014). Les obstacles à l'innovation en France : analyse et recommandations. *Management et Avenir*, 69, 70–88.

Saty Kouame, D. (2012). Le financement, levier de performance des jeunes entreprises innovantes. *Entreprendre et Innover*, 16, 7–17.

Savel, S. (2019). Finance participative : 80 % des montants collectés auprès des particuliers vont aux entreprises. Association financement participatif, *La Tribune* [Online]. Available at: https://www.latribune.fr/entreprises-finance/banques-finance/finance-participative-80-des-montants-collectes-aupres-des-particuliers-vont-aux-entreprises-807148.html.

Savignac, F. (2007). Quel mode de financement pour les jeunes entreprises innovantes : financement interne, prêt bancaire, ou capital-risque ? *Revue économique*, 58(4), 863–889.

Schmitt, G. and Cohen Boulakia, D. (2019). Innovation : lueur ou illumination ? *Décryptages*, 11, 3–7.

Schumpeter, J. (1911). *Théorie de l'évolution économique*. Dalloz, Paris.

Schwienbacher, A. (2008). Innovation and venture capital exits. *The Economic Journal*, 118(533), 1888–1916.

Shi, C. (2003). On the trade-off between future benefits and riskiness of R&D: A bondholder's perspective. *Journal of Accounting and Economics*, 35(2), 227–254.

Simsek, A. (2013a). Financial innovation and portfolio risks. *American Economic Review*, 103, 398–401.

Simsek, A. (2013b). Speculation and risk sharing with new financial assets. *The Quarterly Journal of Economics*, 128, 1365–1396.

Slater, S. and Narver, J. (1998). Customer-led and market-oriented: Let's not confuse the two. *Strategic Management Journal*, 19(10), 1001–1006.

Stewart, T.A. (1997). *Intellectual Capital: The New Wealth of Organizations*. Doubleday, New York.

Stinchcomb, A.L. (1965). Social structure and organization. In *Handbook of Organizations*, March, J.G. (ed.). Rand McNally, Chicago.

Sveiby, K.E. (1997). The intangible asset monitor. *Journal of Human Resource Costing and Accounting*, (2)1, 73–97.

Taleb, N. (2007). *The Black Swan: The Impact of the Highly Improbable*. Penguin, London.

Teece, D.J. (2010). Business models, business strategy and innovation. *Long Range Planning*, 43(2), 172–194.

Teece, D.J. (2011). Dynamic capabilities: A guide for managers. *Ivey Business Journal*, 75(2), 29–32.

Ueda, M. (2004). Banks versus venture capital: Project evaluation, screening, and expropriation. *The Journal of Finance*, 59(2), 601–621.

Valliere, D. (2006). Consequences of growth: Shaping entrepreneurial attitudes. *International Journal of Entrepreneurship and Innovation*, 7(3), 141–148.

Yitmen, I. (2011). Intellectual capital: A competitive asset for driving innovation in engineering design firms. *Engineering Management Journal*, 23(2), 3–19.

Zhang, W. (2015). R&D investment and distress risk. *Journal of Empirical Finance*, 32, 94–114.

Zott, C. and Amit, R. (2002). Measuring the performance implications of business model design: Evidence from emerging growth public firms. Working paper 2002/13/ENT/SM, INSEAD, Fontainebleau.

Index

M, O, P

R, S, T

U, V

Other titles from

ISTE

in

Innovation, Entrepreneurship and Management

2021

BOBILLIER CHAUMON Marc-Eric
Digital Transformations in the Challenge of Activity and Work: Understanding and Supporting Technological Changes
(Technological Changes and Human Resources Set – Volume 3)

2020

ACH Yves-Alain, RMADI-SAÏD Sandra
Financial Information and Brand Value: Reflections, Challenges and Limitations

ANDREOSSO-O'CALLAGHAN Bernadette, DZEVER Sam, JAUSSAUD Jacques, TAYLOR Robert
Sustainable Development and Energy Transition in Europe and Asia
(Innovation and Technology Set – Volume 9)

BEN SLIMANE Sonia, M'HENNI Hatem
Entrepreneurship and Development: Realities and Future Prospects
(Smart Innovation Set – Volume 30)

CHOUTEAU Marianne, FOREST Joëlle, NGUYEN Céline
Innovation for Society: The P.S.I. Approach
(Smart Innovation Set – Volume 28)

CORON Clotilde
Quantifying Human Resources: Uses and Analysis
(Technological Changes and Human Resources Set – Volume 2)

CORON Clotilde, GILBERT Patrick
Technological Change
(Technological Changes and Human Resources Set – Volume 1)

CERDIN Jean-Luc, PERETTI Jean-Marie
The Success of Apprenticeships: Views of Stakeholders on Training and Learning
(Human Resources Management Set – Volume 3)

DELCHET-COCHET Karen
Circular Economy: From Waste Reduction to Value Creation
(Economic Growth Set – Volume 2)

DIDAY Edwin, GUAN Rong, SAPORTA Gilbert, WANG Huiwen
Advances in Data Science
(Big Data, Artificial Intelligence and Data Analysis Set – Volume 4)

DOS SANTOS PAULINO Victor
Innovation Trends in the Space Industry
(Smart Innovation Set – Volume 25)

GASMI Nacer
Corporate Innovation Strategies: Corporate Social Responsibility and Shared Value Creation
(Smart Innovation Set – Volume 33)

GOGLIN Christian
Emotions and Values in Equity Crowdfunding Investment Choices 1: Transdisciplinary Theoretical Approach

GUILHON Bernard
Venture Capital and the Financing of Innovation
(Innovation Between Risk and Reward Set – Volume 6)

LATOUCHE Pascal
Open Innovation: Human Set-up
(Innovation and Technology Set – Volume 10)

LIMA Marcos
Entrepreneurship and Innovation Education: Frameworks and Tools
(Smart Innovation Set – Volume 32)

MACHADO Carolina, DAVIM J. Paulo
Sustainable Management for Managers and Engineers

MAKRIDES Andreas, KARAGRIGORIOU Alex, SKIADAS Christos H.
Data Analysis and Applications 3: Computational, Classification, Financial, Statistical and Stochastic Methods
(Big Data, Artificial Intelligence and Data Analysis Set – Volume 5)
Data Analysis and Applications 4: Financial Data Analysis and Methods
(Big Data, Artificial Intelligence and Data Analysis Set – Volume 6)

MASSOTTE Pierre, CORSI Patrick
Complex Decision-Making in Economy and Finance

MEUNIER François-Xavier
Dual Innovation Systems: Concepts, Tools and Methods
(Smart Innovation Set – Volume 31)

MICHAUD Thomas
Science Fiction and Innovation Design (Innovation in Engineering and Technology Set – Volume 6)

MONINO Jean-Louis
Data Control: Major Challenge for the Digital Society
(Smart Innovation Set – Volume 29)

MORLAT Clément
Sustainable Productive System: Eco-development versus Sustainable Development
(Smart Innovation Set – Volume 26)

SAULAIS Pierre, ERMINE Jean-Louis
Knowledge Management in Innovative Companies 2: Understanding and Deploying a KM Plan within a Learning Organization
(Smart Innovation Set – Volume 27)

2019

AMENDOLA Mario, GAFFARD Jean-Luc
Disorder and Public Concern Around Globalization

BARBAROUX Pierre
Disruptive Technology and Defence Innovation Ecosystems
(Innovation in Engineering and Technology Set – Volume 5)

DOU Henri, JUILLET Alain, CLERC Philippe
Strategic Intelligence for the Future 1: A New Strategic and Operational Approach
Strategic Intelligence for the Future 2: A New Information Function Approach

FRIKHA Azza
Measurement in Marketing: Operationalization of Latent Constructs

FRIMOUSSE Soufyane
Innovation and Agility in the Digital Age
(Human Resources Management Set – Volume 2)

GAY Claudine, SZOSTAK Bérangère L.
Innovation and Creativity in SMEs: Challenges, Evolutions and Prospects
(Smart Innovation Set – Volume 21)

GORIA Stéphane, HUMBERT Pierre, ROUSSEL Benoît
Information, Knowledge and Agile Creativity
(Smart Innovation Set – Volume 22)

HELLER David
Investment Decision-making Using Optional Models
(Economic Growth Set – Volume 2)

HELLER David, DE CHADIRAC Sylvain, HALAOUI Lana, JOUVET Camille
The Emergence of Start-ups
(Economic Growth Set – Volume 1)

HÉRAUD Jean-Alain, KERR Fiona, BURGER-HELMCHEN Thierry
Creative Management of Complex Systems
(Smart Innovation Set – Volume 19)

LATOUCHE Pascal
Open Innovation: Corporate Incubator
(Innovation and Technology Set – Volume 7)

LEHMANN Paul-Jacques
The Future of the Euro Currency

LEIGNEL Jean-Louis, MÉNAGER Emmanuel, YABLONSKY Serge
Sustainable Enterprise Performance: A Comprehensive Evaluation Method

LIÈVRE Pascal, AUBRY Monique, GAREL Gilles
Management of Extreme Situations: From Polar Expeditions to Exploration-Oriented Organizations

MILLOT Michel
Embarrassment of Product Choices 2: Towards a Society of Well-being

N'GOALA Gilles, PEZ-PÉRARD Virginie, PRIM-ALLAZ Isabelle
Augmented Customer Strategy: CRM in the Digital Age

NIKOLOVA Blagovesta
The RRI Challenge: Responsibilization in a State of Tension with Market Regulation
(Innovation and Responsibility Set – Volume 3)

PELLEGRIN-BOUCHER Estelle, ROY Pierre
Innovation in the Cultural and Creative Industries
(Innovation and Technology Set – Volume 8)

PRIOLON Joël
Financial Markets for Commodities

QUINIOU Matthieu
Blockchain: The Advent of Disintermediation

RAVIX Joël-Thomas, DESCHAMPS Marc
Innovation and Industrial Policies
(Innovation between Risk and Reward Set – Volume 5)

ROGER Alain, VINOT Didier
Skills Management: New Applications, New Questions
(Human Resources Management Set – Volume 1)

SAULAIS Pierre, ERMINE Jean-Louis
Knowledge Management in Innovative Companies 1: Understanding and Deploying a KM Plan within a Learning Organization
(Smart Innovation Set – Volume 23)

SERVAJEAN-HILST Romaric
Co-innovation Dynamics: The Management of Client-Supplier Interactions for Open Innovation
(Smart Innovation Set – Volume 20)

SKIADAS Christos H., BOZEMAN James R.
Data Analysis and Applications 1: Clustering and Regression, Modeling-estimating, Forecasting and Data Mining
(Big Data, Artificial Intelligence and Data Analysis Set – Volume 2)
Data Analysis and Applications 2: Utilization of Results in Europe and Other Topics
(Big Data, Artificial Intelligence and Data Analysis Set – Volume 3)

UZUNIDIS Dimitri
Systemic Innovation: Entrepreneurial Strategies and Market Dynamics

VIGEZZI Michel
World Industrialization: Shared Inventions, Competitive Innovations and Social Dynamics
(Smart Innovation Set – Volume 24)

2018

BURKHARDT Kirsten
Private Equity Firms: Their Role in the Formation of Strategic Alliances

CALLENS Stéphane
Creative Globalization
(Smart Innovation Set – Volume 16)

CASADELLA Vanessa
Innovation Systems in Emerging Economies: MINT – Mexico, Indonesia, Nigeria, Turkey
(Smart Innovation Set – Volume 18)

CHOUTEAU Marianne, FOREST Joëlle, NGUYEN Céline
Science, Technology and Innovation Culture
(Innovation in Engineering and Technology Set – Volume 3)

CORLOSQUET-HABART Marine, JANSSEN Jacques
Big Data for Insurance Companies
(Big Data, Artificial Intelligence and Data Analysis Set – Volume 1)

CROS Françoise
Innovation and Society
(Smart Innovation Set – Volume 15)

DEBREF Romain
Environmental Innovation and Ecodesign: Certainties and Controversies
(Smart Innovation Set – Volume 17)

DOMINGUEZ Noémie
SME Internationalization Strategies: Innovation to Conquer New Markets

ERMINE Jean-Louis
Knowledge Management: The Creative Loop
(Innovation and Technology Set – Volume 5)

GILBERT Patrick, BOBADILLA Natalia, GASTALDI Lise,
LE BOULAIRE Martine, LELEBINA Olga
Innovation, Research and Development Management

IBRAHIMI Mohammed
Mergers & Acquisitions: Theory, Strategy, Finance

LEMAÎTRE Denis
Training Engineers for Innovation

LÉVY Aldo, BEN BOUHENI Faten, AMMI Chantal
Financial Management: USGAAP and IFRS Standards
(Innovation and Technology Set – Volume 6)

MILLOT Michel
Embarrassment of Product Choices 1: How to Consume Differently

PANSERA Mario, OWEN Richard
Innovation and Development: The Politics at the Bottom of the Pyramid
(Innovation and Responsibility Set – Volume 2)

RICHEZ Yves
Corporate Talent Detection and Development

SACHETTI Philippe, ZUPPINGER Thibaud
New Technologies and Branding
(Innovation and Technology Set – Volume 4)

SAMIER Henri
Intuition, Creativity, Innovation

TEMPLE Ludovic, COMPAORÉ SAWADOGO Eveline M.F.W.
Innovation Processes in Agro-Ecological Transitions in Developing Countries
(Innovation in Engineering and Technology Set – Volume 2)

UZUNIDIS Dimitri
Collective Innovation Processes: Principles and Practices
(Innovation in Engineering and Technology Set – Volume 4)

VAN HOOREBEKE Delphine
The Management of Living Beings or Emo-management

2017

AÏT-EL-HADJ Smaïl
The Ongoing Technological System
(Smart Innovation Set – Volume 11)

BAUDRY Marc, DUMONT Béatrice
Patents: Prompting or Restricting Innovation?
(Smart Innovation Set – Volume 12)

BÉRARD Céline, TEYSSIER Christine
Risk Management: Lever for SME Development and Stakeholder Value Creation

CHALENÇON Ludivine
Location Strategies and Value Creation of International Mergers and Acquisitions

CHAUVEL Danièle, BORZILLO Stefano
The Innovative Company: An Ill-defined Object
(Innovation between Risk and Reward Set – Volume 1)

CORSI Patrick
Going Past Limits To Growth

D'ANDRIA Aude, GABARRET Inés
Building 21st Century Entrepreneurship
(Innovation and Technology Set – Volume 2)

DAIDJ Nabyla
Cooperation, Coopetition and Innovation
(Innovation and Technology Set – Volume 3)

FERNEZ-WALCH Sandrine
The Multiple Facets of Innovation Project Management
(Innovation between Risk and Reward Set – Volume 4)

FOREST Joëlle
Creative Rationality and Innovation
(Smart Innovation Set – Volume 14)

GUILHON Bernard
Innovation and Production Ecosystems
(Innovation between Risk and Reward Set – Volume 2)

HAMMOUDI Abdelhakim, DAIDJ Nabyla
Game Theory Approach to Managerial Strategies and Value Creation
(Diverse and Global Perspectives on Value Creation Set – Volume 3)

LALLEMENT Rémi
Intellectual Property and Innovation Protection: New Practices and New Policy Issues
(Innovation between Risk and Reward Set – Volume 3)

LAPERCHE Blandine
Enterprise Knowledge Capital
(Smart Innovation Set – Volume 13)

LEBERT Didier, EL YOUNSI Hafida
International Specialization Dynamics
(Smart Innovation Set – Volume 9)

MAESSCHALCK Marc
Reflexive Governance for Research and Innovative Knowledge
(Responsible Research and Innovation Set – Volume 6)

MASSOTTE Pierre
Ethics in Social Networking and Business 1: Theory, Practice and Current Recommendations
Ethics in Social Networking and Business 2: The Future and Changing Paradigms

MASSOTTE Pierre, CORSI Patrick
Smart Decisions in Complex Systems

MEDINA Mercedes, HERRERO Mónica, URGELLÉS Alicia
Current and Emerging Issues in the Audiovisual Industry
(Diverse and Global Perspectives on Value Creation Set – Volume 1)

MICHAUD Thomas
Innovation, Between Science and Science Fiction
(Smart Innovation Set – Volume 10)

PELLÉ Sophie
Business, Innovation and Responsibility
(Responsible Research and Innovation Set – Volume 7)

SAVIGNAC Emmanuelle
The Gamification of Work: The Use of Games in the Workplace

SUGAHARA Satoshi, DAIDJ Nabyla, USHIO Sumitaka
Value Creation in Management Accounting and Strategic Management: An Integrated Approach
(Diverse and Global Perspectives on Value Creation Set –Volume 2)

UZUNIDIS Dimitri, SAULAIS Pierre
Innovation Engines: Entrepreneurs and Enterprises in a Turbulent World
(Innovation in Engineering and Technology Set – Volume 1)

2016

BARBAROUX Pierre, ATTOUR Amel, SCHENK Eric
Knowledge Management and Innovation
(Smart Innovation Set – Volume 6)

BEN BOUHENI Faten, AMMI Chantal, LEVY Aldo
Banking Governance, Performance And Risk-Taking: Conventional Banks Vs Islamic Banks

BOUTILLIER Sophie, CARRÉ Denis, LEVRATTO Nadine
Entrepreneurial Ecosystems (Smart Innovation Set – Volume 2)

BOUTILLIER Sophie, UZUNIDIS Dimitri
The Entrepreneur (Smart Innovation Set – Volume 8)

BOUVARD Patricia, SUZANNE Hervé
Collective Intelligence Development in Business

GALLAUD Delphine, LAPERCHE Blandine
Circular Economy, Industrial Ecology and Short Supply Chains
(Smart Innovation Set – Volume 4)

GUERRIER Claudine
Security and Privacy in the Digital Era
(Innovation and Technology Set – Volume 1)

MEGHOUAR Hicham
Corporate Takeover Targets

MONINO Jean-Louis, SEDKAOUI Soraya
Big Data, Open Data and Data Development
(Smart Innovation Set – Volume 3)

MOREL Laure, LE ROUX Serge
Fab Labs: Innovative User
(Smart Innovation Set – Volume 5)

PICARD Fabienne, TANGUY Corinne
Innovations and Techno-ecological Transition
(Smart Innovation Set – Volume 7)

2015

CASADELLA Vanessa, LIU Zeting, DIMITRI Uzunidis
Innovation Capabilities and Economic Development in Open Economies
(Smart Innovation Set – Volume 1)

CORSI Patrick, MORIN Dominique
Sequencing Apple's DNA

CORSI Patrick, NEAU Erwan
Innovation Capability Maturity Model

FAIVRE-TAVIGNOT Bénédicte
Social Business and Base of the Pyramid

GODÉ Cécile
Team Coordination in Extreme Environments

MAILLARD Pierre
Competitive Quality and Innovation

MASSOTTE Pierre, CORSI Patrick
Operationalizing Sustainability

MASSOTTE Pierre, CORSI Patrick
Sustainability Calling

2014

DUBÉ Jean, LEGROS Diègo
Spatial Econometrics Using Microdata

LESCA Humbert, LESCA Nicolas
Strategic Decisions and Weak Signals

2013

HABART-CORLOSQUET Marine, JANSSEN Jacques, MANCA Raimondo
VaR Methodology for Non-Gaussian Finance

2012

DAL PONT Jean-Pierre
Process Engineering and Industrial Management

MAILLARD Pierre
Competitive Quality Strategies

POMEROL Jean-Charles
Decision-Making and Action

SZYLAR Christian
UCITS Handbook

2011

LESCA Nicolas
Environmental Scanning and Sustainable Development

LESCA Nicolas, LESCA Humbert
Weak Signals for Strategic Intelligence: Anticipation Tool for Managers

MERCIER-LAURENT Eunika
Innovation Ecosystems

2010

SZYLAR Christian
Risk Management under UCITS III/IV

2009

COHEN Corine
Business Intelligence

ZANINETTI Jean-Marc
Sustainable Development in the USA

2008

CORSI Patrick, DULIEU Mike
The Marketing of Technology Intensive Products and Services

DZEVER Sam, JAUSSAUD Jacques, ANDREOSSO Bernadette
Evolving Corporate Structures and Cultures in Asia: Impact of Globalization

2007

AMMI Chantal
Global Consumer Behavior

2006

BOUGHZALA Imed, ERMINE Jean-Louis
Trends in Enterprise Knowledge Management

CORSI Patrick *et al.*
Innovation Engineering: the Power of Intangible Networks